IoT Fundamentals with a Practical Approach

IoT Fundamentals with a Practical Approach is an insightful book that serves as a comprehensive guide to understanding the foundations and key concepts of Internet of Things (IoT) technologies.

The book begins by introducing readers to the concept of IoT, explaining the significance and potential impact on various industries and domains. It covers the underlying principles of IoT, including its architecture, connectivity, and communication protocols, providing readers with a solid understanding of how IoT systems are structured and how devices interact within an IoT ecosystem.

This book dives into the crucial components that form the backbone of IoT systems. It explores sensors and actuators, explaining their roles in collecting and transmitting data from the physical environment. The book also covers electronic components used in IoT devices, such as microcontrollers, communication modules, and power management circuits. This comprehensive understanding of the building blocks of IoT allows readers to grasp the technical aspects involved in developing IoT solutions.

Security is a vital aspect of IoT, and the book dedicates a significant portion to exploring security challenges and best practices in IoT deployments. It delves into topics such as authentication, encryption, access control, and secure firmware updates, providing readers with essential insights into safeguarding IoT systems against potential threats and vulnerabilities.

This book also addresses the scalability and interoperability challenges of IoT. It discusses IoT platforms and frameworks that facilitate the development and management of IoT applications, highlighting their role in enabling seamless integration and communication between devices and systems.

The book is written in a clear and accessible manner and includes real-world examples, making it suitable for both beginners and professionals looking to enhance their understanding of IoT. It serves as a valuable resource for engineers, developers, researchers, and decision-makers involved in IoT projects and provides them with the knowledge and tools necessary to design, implement, and secure IoT solutions.

IoT Fundamentals with a Practical Approach

Neera Batra
Sonali Goyal

CRC Press
Taylor & Francis Group
Boca Raton London New York

CRC Press is an imprint of the
Taylor & Francis Group, an **informa** business

Designed cover image: © Shutterstock

First edition published 2025
by CRC Press
2385 NW Executive Center Drive, Suite 320, Boca Raton FL 33431

and by CRC Press
4 Park Square, Milton Park, Abingdon, Oxon, OX14 4RN

CRC Press is an imprint of Taylor & Francis Group, LLC

© 2025 Neera Batra and Sonali Goyal

ISBN: 978-1-032-30969-9 (hbk)
ISBN: 978-1-032-30970-5 (pbk)
ISBN: 978-1-003-30748-8 (ebk)

DOI: 10.1201/9781003307488

Typeset in Sabon
by SPi Technologies India Pvt Ltd (Straive)

Contents

About the Authors

Dr. Neera Batra is an accomplished scholar with a robust academic background. In 2012, she earned her Doctor of Philosophy (PhD) in Computer Science & Engineering from Maharishi Markandeshwar (deemed to be University), Mullana, Ambala, India. Prior to this, she obtained her Master of Technology (MTech.) in Computer Science & Engineering from Kurukshetra University, Kurukshetra, India, in 2007.

Since 2007, Dr. Neera Batra has been actively engaged in both teaching and research and development. Her dedication to academia is evident through her supervision of numerous MTech. and PhD theses, contributing significantly to the academic and professional growth of her students.

Dr. Batra's scholarly contributions also extend beyond the classroom. She has authored and published over 70 research papers in prominent national and international journals, in addition to presenting her work at various refereed national and international conferences. Her commitment to innovation is further exemplified by her impressive portfolio of 16 patented inventions.

Dr. Batra's research interests are multifaceted, with a focus on cutting-edge areas such as the Internet of Things (IoT), machine learning, and software engineering. Her valuable contributions to these fields demonstrate her expertise and dedication to advancing the frontier of knowledge.

Dr. Sonali Goyal is an accomplished academic and researcher in the field of machine learning and the Internet of Things. As an associate professor in the Department of CSE, MMEC, Maharishi Markandeshwar (deemed to be University), Mullana, she has made significant contributions to the advancement of knowledge in her field. With a strong background in both theory and practical applications, she has dedicated herself to exploring the potential of cutting-edge technologies. Her academic career spans over 12 years, during which she has been actively engaged in teaching, mentoring and conducting research. In terms of research, Dr. Sonali Goyal has a prolific record, having authored 25 research papers on various aspects of machine learning and the Internet of Things. Her work demonstrates a deep understanding of the underlying principles and a keen ability to apply them to practical problems. In addition to her research papers, she has also obtained five patents for her innovative contributions to the field. She has also authored a book, showcasing her ability to communicate complex concepts effectively.

Chapter 1

Introduction to IoT (Internet of Things)

1.1 INTRODUCTION TO THE INTERNET

The Internet is an increasingly important part of everyday life for people all around the world. It is the foremost important tool and the most prominent resource, being used by almost every person across the globe. The Internet is a worldwide network of billions of computers and other electronic devices. Through its use, it's possible to access almost any information, communicate with anyone else in the world, and do much more. In other words, the Internet is a widespread interconnected network of computers and electronic devices (which support the system). It creates a communication medium to share and obtain information online. If your device is connected to the Internet then you will be able to access all the applications, websites, social media apps, and many more services.

At present, the Internet is considered to be the fastest medium for sending and receiving information. It connects millions of computers, web pages, websites, and servers. Using the Internet you can send emails, photos, videos, and messages to our loved ones. You can do all of this by connecting a computer to the Internet, which is also known as 'going online'. Overview of IoT (Internet of Things).

Humans' desire for a comfortable living develops from their curiosity about the mechanical world. Over the last few decades, humankind had experienced a transformational technological journey which has crossed many new frontiers. These frontiers have interacted with human beings and performed every possible work in a shorter period of time and with a much greater degree of accuracy. With the advent of 'smart concepts', the world is now becoming increasingly connected. A more accurate term for it might be a hyper-connected world. These smart concepts include smartphones, smart devices, smart applications, and smart cities. These smarter concepts form an ecosystem of devices whose basic work is to connect various devices to send and receive data. The Internet of Things is the one dominant technology which keeps an eye on connected smart devices. The Internet of Things has bought applications from fiction to fact, thereby enabling the fourth industrial revolution. It has laid an incredible impact on the technical, social, economic factors and also on the lives of human and machines. Scientists claim that the potential benefits to be derived from this technology will produce a future in which the smart objects sense, think, and act. The IoT is the trending technology and embodies various related concepts such as fog computing, edge computing, communication protocols, electronic devices, sensors, geolocation etc. The chapter presents comprehensive information about the evolution of Internet of Things, moving from its present developments to its futuristic applications.

With the advent of the latest techniques and technologies, there was a need for a concept that would describe how the Internet would expand as sensors and intelligence are added to physical items such as consumer devices or physical assets and these objects are connected to the Internet. The vision and concept have existed for years; however, there has been an acceleration in the number and types of things that were required to be being connected and used in the technologies for identifying, sensing, and communicating. Here comes the technology's "IoT" to the rescue.

There is presently a great deal of noise at the moment about the IoT and its future impact on everything from the way we travel and do our shopping to the way in which manufacturers keep track of inventory. But what is the Internet of Things? How does it work? And is it really that important?

1.1.1 What is the Internet of Things (IoT)?

The term Internet of Things, or IoT, is the name given to the collective network of connected devices and the technology that facilitates communication between devices and the cloud, as well as between the devices themselves. Thanks to the invention of inexpensive computer chips and the spread of high bandwidth telecommunication, we now have billions of devices connected to the Internet. This means everyday devices such as toothbrushes, vacuums, cars, and a wide range of machines can use sensors to collect data and respond intelligently to users.

The IoT integrates everyday "things" with the Internet. Computer engineers have been adding sensors and processors to everyday objects since the 1990s. However, progress was initially slow because the chips were big and bulky. Low-power computer chips, called Radio Frequency Identification (RFID) tags, were first used to track expensive equipment. As computing devices shrank in size, these chips also became smaller, faster, and smarter over time.

The IoT is presently a hot technology across the globe. Government, academia, and industry are involved in different aspects of research, implementation, and business with IoT. The technology cuts across different application domain verticals, ranging from civilian to defence sectors. Among the sectors in which the IoT is being disseminated are agriculture, space, healthcare, manufacturing, construction, water, and mining, which are presently transitioning their legacy infrastructure to support IoT. Today it is possible to envision pervasive connectivity, storage, and computation, which, in turn, gives rise to the construction of different IoT solutions. IoT-based applications such as an innovative shopping system, infrastructure management in both urban and rural areas, remote health monitoring and emergency notification systems, and transportation systems, are gradually increasing their reliance on IoT-based systems. Therefore, it is very important to learn the fundamentals of this emerging technology.

IoT includes an extraordinary number of objects of all shapes and sizes – from smart microwaves, which automatically cook your food for the right length of time, to self-driving cars, whose complex sensors detect objects in their path, to wearable fitness devices that measure your heart rate and the number of steps you've taken that day, then use that information to suggest exercise plans tailored to you. There are even connected footballs that can track how far and fast they are thrown and record those statistics via an app for future training purposes.

Let's start with a simple real-life example: Rajesh, in between his road trips, notices some problem with the check engine light. However, he is unaware of the intensity of the problem. The good part is that the sensor that triggers the check engine light monitors the pressure in the inner brake line. This sensor is one of the many sensors present in the car which constantly communicate with each other. A component, called the diagnostic bus, gathers the data from all these sensors and then passes it to the gateway in the car. The gateway collects and sorts the data from different sensors.

Before this connection can take place, the car's gateway and platform must register with each other and confirm a secure communication connection. The platform keeps on constantly gathering and storing information from hundreds of cars worldwide, building a record in a database. The manufacturer has added rules and logic to the platform. The platform triggers an alert in his car, after sensing the brake fluid has dropped below the recommended level. The manufacturer then sends him an appointment for servicing of his car, and the car's problem is rectified.

1.1.2 What are the things?

Things are objects either of the physical world (physical things) or of the information world (the virtual world) which are capable of being identified and integrated into communication networks. Physical things exist in the physical world and are capable of being sensed, actuated, and connected. These include, for example, the surrounding environment, industrial robots, goods, and electrical equipment. Virtual things, by contrast, exist in the information world and are capable of being stored, processed, and accessed. Example of these include multimedia content and application software. The thing in the IoT could also be alive, as would be the case, for example, with a person with a diabetes monitor implant or an animal with a tracking device.

1.1.3 Internet of Things vs the Internet

The main difference between the Internet of Things and the Internet is the identity of the content creator. On the conventional Internet, content is consumed on a request basis. In the IoT, on the other hand, the material is often consumed by sending a notice or initiating an action when a condition of interest is discovered.

As stated earlier, the IoT is a network of physical items that is infused with technology and linked to the Internet, as well as to other connected devices. These items capture and transmit information about how they're utilized and their surroundings. Status data, automation data, and location data are three forms of IoT data that vary depending on the device that generates it and the case study involved. The Internet is a vast network that connects numerous computers and other electronic gadgets all around the world. Anyone can get nearly any information, interact with anyone on the globe, and do a lot more using the Internet. Decentralization is a feature of the internet. Nobody possesses the Internet or has control over who can access it.

1.1.4 Comparison table between the Internet of Things and the Internet (Table 1.1)

Table 1.1 The Internet of Things (IoT) vs the Internet

Parameters of comparison	Internet of things	Internet
Objective	Focused on the actual world.	Focused more on the virtual world.
Tasks that are done so far	Content creation.	Content generation and consumption.
Based on	Concepts of physical-first.	Concepts of physical-first and digital-first.
Connection type	Multipoint.	Point-to-point as well as multipoint.
Content combined with	Explicitly defined operators.	Physical linkages.

1.1.5 Main differences between the Internet of Things and the Internet

1. With the Internet of Things, increased attention is placed on the actual world rather than the virtual world, promoting a better balance of virtual and real experiences. This is in contrast to the Internet, which is more inclined towards the virtual world as shown in above Table 1.1.

- The IoT is primarily concerned with content creation, whereas the Internet is concerned with both content generation and consumption.
- The IoT is based on the physical-first notion, whereas the Internet is based on both the physical-first and digital-first concepts.
- IoT uses a multipoint connection, whereas the Internet uses both point-to-point and multipoint connections.
- Physical linkages between websites are used to connect users on the conventional Internet whereas in the Internet of Things the content is merged using operators that are expressly stated.

The IoT is a collection of interconnected devices. As a result, higher integration abilities and end-to-end thinking are required for the Internet of Things. Smart homes and smart cities, as well as manufacturing, telemedicine, and precision agriculture, are all being advanced by the Internet of Things.

While the full promise of the Internet of Things has yet to be realized, it already has a wide range of practical uses in the real world. The Internet is a global network that allows enterprises, governments, colleges, and other institutions to communicate with one another. The result is a maze of wires, computer systems, storage systems, routers, servers, repeaters, satellites, and Wi-Fi towers that allow digital data to go around the globe.

1.1.6 Goals of IoT

The goals of the IoT are to extend to Internet connectivity from standard devices such as computers, mobile phones, and tablets to relatively "dumb" devices such as a toaster. IoT makes virtually everything "smart," by improving aspects of our life with the power of data collection, AI algorithm, and networks.

1.1.7 Origin of the IoT

Development and historical background: The idea of adding intelligence via sensors and other hardware components to physical devices in order to enable connectivity between them has been debated since the 1980s. At that time, however, we were only able to go as far as Internet-connected vending machines. The limitations at this time were expensive components, bulkier computer chips, and the inconsistent Internet signal. The introduction and adoption of RFID tags helped to curb the issue to a certain extent. In addition, the adoption of IPv6 helped the idea to progress further.

The expression Internet of Things was first used in 1999 by Kevin Ashton, the executive director of Auto-ID labs at MIT, while he was giving a presentation for Procter & Gamble. In his presentation, he noted how in today's computer, i.e. the Internet at that time, the computer was dependent on personal input. Almost all the data collected on the Internet was captured by people either typing, pressing a record button, or scanning a barcode. People are incapable of capturing all the data in the world and if we have computers that can inherit all the data without any input from us, this would reduce both the levels of waste and the costs involved. We would know when things need to be repaired or replaced considering the best for everyone.

Before 1999, the phrase was referred to as Radio Frequency Identification (RFID), which was used for tracking consignments. One of the first examples of the IoT was the installation of the Coca-Cola machine at Carnegie Melon University in the 1980s, which meant that local programmers would connect the machine with the Internet and check the availability and temperature of the Coca-Cola, checking if it is cold enough before physically going to take it out from the refrigerator. Kevin Ashton believed that if all the devices are designated in such a way the computer could manage to track the databases and inventory them. In 1999 he coined the term 'Internet of Things', but it took the technology at least another decade to catch up with the idea. In one of the first applications of the IoT, RFID tags were added to types of equipment to track their location. Following this, prices have been falling for the sensors, hardware, and Internet connections, paving the way to connect possibly everything to the Internet. Slowly and steadily, IoT has started to spread from manufacturing and businesses to homes and offices. At present, most of the areas are possibly connected to the Internet, allowing the IoT to emerge as one of the most essential technologies in the coming years.

As of now, devices are designed in a way to track them from anywhere. The IoT provides an ample supply of opportunities to interconnect our devices and equipment. Inventory control is one of the prominent advantages of the Internet of Things.

1.1.8 Evolution of IoT

- 1970 – The actual idea of connected devices was proposed
- 1990 – John Romkey created a toaster which could be turned on/off over the network
- 1995 – Siemens introduced the first cellular module built for machine-to-machine (M2M) communication
- 1999 – The term "Internet of Things" was used by Kevin Ashton during his work at P&G which became widely accepted
- 2004 – The term was mentioned in famous publications like the *Guardian*, the *Boston Globe*, and *Scientific American*
- 2005 – UN's International Telecommunications Union (ITU) published its first report on this topic.
- 2008 – The Internet of Things was born

- 2011 – Gartner, the market research company, include "The Internet of Things" technology in their research
- On 15 October 2015 the Internet Society published this 50-page whitepaper, providing an overview of the IoT and exploring related issues and challenges.
- 22 Oct 2015 – PDF file updated with higher-quality cover image and a title page.
- 6 Jan 2016 – PDF file updated with new graphic design. Filename changed to include "-en" for English.
- 18 Apr 2016 – Russian translation published.
- 17 Aug 2016 – Spanish translation published.

The Internet of Things is an emerging topic of technical, social, and economic significance. Consumer products, durable goods, cars and trucks, industrial and utility components, sensors, and other everyday objects are being combined with Internet connectivity and powerful data analytic capabilities that promise to transform the way in which we work, live, and play. Projections for the impact of IoT on the Internet and the wider global economy are impressive, with some anticipating that by 2025 there will be as many as 100 billion connected IoT devices and a global economic impact of more than $11 trillion.

At the same time, however, the Internet of Things raises significant challenges that could stand in the way of realizing its potential benefits. Attention-grabbing headlines about the hacking of Internet-connected devices, surveillance concerns, and privacy fears have already raised public concerns. Technical challenges remain and new policy, legal, and development changes are emerging. As a matter of principle, developers and users of IoT devices and systems have a collective obligation to ensure they do not expose users and the Internet itself to potential harm. Accordingly, a collaborative approach to security will be needed to develop effective and appropriate solutions to IoT security challenges that are well suited to the scale and complexity of the issues.

1.1.9 The history of the Internet of Things

The idea of devices exchanging information without human appeared not long ago. Full automation of data transmission was discussed in the late 1970s. At that time this approach was considered as "pervasive computing". It took several decades for technologies' development to start talking about the Internet of Things.

As mentioned above, In the second half of the 1990s, the Briton Kevin Ashton was working for Procter & Gamble enterprises and was engaged in the optimization of the production process. He noticed that this process directly depends on the speed of transmission and processing of data. It can take days for people who collect the data. The use of RFID has allowed accelerating the process of data transfer directly between devices. He had an idea of things to be collected, processed and transmitted with no human involvement. He decided to call it an "Internet of Things" and became a visionary at that time.

It took almost a decade for the "Internet of Things" term to come into common use in everyday life. Together with artificial intelligence, IoT has become a cutting edge in the development of information technology. So, in 2008 IPSO Alliance created an alliance of companies that supported the development of the Internet of Things technologies. It has become a signal for large corporations.

In the summer of 2010, it became known that the Google Street View service was not only showing panoramic photos, but was also able to collect data through the use of Wi-Fi. Experts talked about the development of a new protocol for data transmission, which would allow the free exchange of data between devices. In the same year, China announced that it was planning to include the Internet on the list of priority research areas for the next five years. It became clear that not only large corporations but also the government were interested in collecting, processing, and storing data. In 2011, Gartner, a market research firm, named the IoT in its list of the most promising emerging technologies.

Over the next few years the Internet of Things essentially conquered the world. In 2012, the largest European Internet conference Le Web was devoted to this topic. Many of the key business magazines, such as *Forbes*, *Fast Company, and Wired*, began to make active use of the term. The whole world started to discuss the Internet of Things while the companies launched the Internet of Things technology race. In 2013, IDC published a study that predicted the growth of the IoT market by 2020 to $8.9 trillion.

In January 2014, Google bought a $3.2 million company that was developing smart home appliances and building management systems. Since then, the world market has fully understood that the nearest future belongs to the Internet of Things. In the same year, the globally important American Consumer Electronics Show was held in Las Vegas under the heading of the Internet of Things.

1.2 HOW DOES IoT WORK?

Devices and objects with built-in sensors are connected to an Internet of Things platform, which integrates data from the different devices and applies analytics to share the most valuable information with applications built to address specific needs.

These powerful IoT platforms can pinpoint exactly what information is useful and what can safely be ignored. This information can be used to detect patterns, make recommendations, and detect possible problems before they occur. Figure 1.1 shows life before and after IoT.

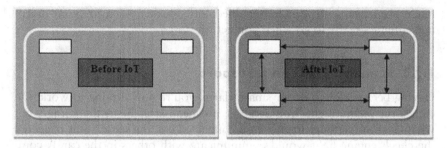

Figure 1.1 Before and After IoT.

For example, if I own a car manufacturing business, I might want to know which optional components (leather seats or alloy wheels, for example) are the most popular. Using Internet of Things technology, I can:

- Use sensors to detect which areas in a showroom are the most popular, and where customers linger longest;
- Drill down into the available sales data to identify which components are selling fastest;
- Automatically align sales data with supply, so that popular items don't go out of stock.

The information picked up by connected devices enables people to make smart decisions about which components to stock up on, based on real-time information, which helps them to save time and money.

With the insights provided by advanced analytics comes the power to make processes more efficient. Smart objects and systems mean you can automate certain tasks, particularly ones which are repetitive, mundane, time-consuming, or even dangerous.

Let's look at some examples to see what this looks like in real life.

1.2.1 Scenario 1: IoT in your home

Imagine you wake up at 7a.m. every day to go to work. Your alarm clock does the job of waking you just fine. That is, until something goes wrong. Your train's cancelled and you have to drive to work instead. The only problem is that it takes longer to drive, and you would have needed to get up at 6.45a.m. to avoid being late. Oh, and it's also pouring with rain, so you'll need to drive slower than usual. A connected or IoT-enabled alarm clock would reset itself based on all these factors, to ensure you got to work on time. It could recognize that your usual train is cancelled, calculate the driving distance and travel time for your alternative route to work, check the weather and factor in slower travelling speed because of heavy rain, and calculate when it needs to wake you up so you're not late. If it's super-smart, if might even sync with your IoT-enabled coffee maker, to ensure your morning caffeine's ready to go when you get up.

1.2.2 Scenario 2: IoT in transport

Having been woken by your smart alarm, you're now driving to work. On comes the engine light. You'd rather not head straight to the garage, but what if it's something urgent? In a connected car, the sensor that triggered the check engine light would communicate with others in the car. A component called the diagnostic bus collects data from these sensors and passes it to a gateway in the car, which sends the most relevant information to the manufacturer's platform. The manufacturer can use data from the car to offer you an appointment to get the part fixed, send you directions to the nearest dealer, and make sure the correct replacement part is ordered so it's ready for you when you show up.

The cost of integrating computing power into small objects has now dropped considerably. For example, you can add connectivity with Alexa voice services capabilities to microcontrollers with less than 1 MB embedded RAM, such as for light switches. A whole industry has sprung up with a focus on filling our homes, businesses, and offices with IoT devices. These smart objects can automatically transmit data to and from the Internet. All these "invisible computing devices" and the technology associated with them are collectively referred to as the Internet of Things.

1.3 IoT COMPONENTS

Below, we list four fundamental components of the IoT system, which tells us how it works:

1. **Sensors/Devices**: Sensors or devices are key component that help you to collect live data from the surrounding environment. All this data may have various levels of complexities. It could be a simple temperature monitoring sensor, or it may be in the form of the video feed.

   A device may have various types of sensors which performs multiple tasks **apart** from sensing. For example, a mobile phone is a device which has multiple sensors like GPS and a camera but your smartphone is unable to sense these things.

2. **Connectivity**: All the collected data is sent to a cloud infrastructure. The sensors should be connected to the cloud using various mediums of communications. These communication mediums include mobile or satellite networks, Bluetooth, Wi-Fi, WAN, etc.

3. **Data Processing**: Once that data is collected, and it gets to the cloud, the software performs processing on the gathered data. This process

can be just checking the temperature, reading on devices like AC or heaters. However, it can sometimes also be very complex, such as identifying objects, using computer vision on video.

4. **User Interface**: The information needs to be available to the end-user in some way which can be achieved by triggering alarms on their phones or sending them notification through email or text message. The user sometimes might need an interface which actively checks their IoT system. For example, the user has a camera installed in his home. He wants to access video recording and all the feeds with the help of a web server.

Before we understand the impact IoT can have on our way of living, it's important to consider its advantages and disadvantages:

1.4 ADVANTAGES OF IoT

There are a number of advantages of IoT:

1. **Communication**: IoT encourages the communication between devices, also famously known as machine-to-machine (M2M) communication. Because of this, the physical devices are able to stay connected and hence the total transparency is available with lesser inefficiencies and greater quality.
2. **Automation and control**: Due to physical objects getting connected and controlled digitally and centrally with wireless infrastructure, there is a large amount of automation and control in the workings. Without human intervention, the machines are able to communicate with each other leading to faster and timely output.
3. **Information**: It is obvious that having more information helps making better decisions. Whether it is mundane decisions as needing to know what to buy at the grocery store or if your company has enough widgets and supplies, knowledge is power and more knowledge is better.
4. **Monitor**: The second most obvious advantage of IoT is monitoring. Knowing the exact quantity of supplies or the air quality in your home, can further provide more information that could not have previously been collected easily. For instance, knowing that you are low on milk or on printer ink could spare you another trip to the store in the near future. Furthermore, a monitoring of the expiry dates of products can and will improve safety.
5. **Time**: As hinted in the previous examples, the amount of time saved because of IoT could be quite large. And in today's modern life, we all could use more time.

6. **Money**: The biggest advantage of IoT is saving money. If the price of the tagging and monitoring equipment is less than the amount of money saved, then the Internet of Things will be adopted widely. IoT fundamentally proves to be very helpful to people in their daily routines by making the appliances communicate to each other in an effective manner, thereby saving and conserving energy and cost. Allowing the data to be communicated and shared between devices and then translating it into our required way, it makes our systems effective.

7. **Automation of daily tasks leads to the better monitoring of devices**: The IoT allows you to automate and control the tasks that are carried out on a daily basis, avoiding human intervention. M2M communication helps to maintain transparency in the processes. It also leads to uniformity in the tasks. It can also maintain the quality of service. We can also take necessary action in case of emergencies.

8. **Efficient and saves time**: M2M interaction provides better efficiency, meaning that accurate results can be obtained quickly. This results in the saving of valuable time. Instead of repeating the same tasks every day, it enables people to do other creative jobs.

9. **Saves money**: The optimum utilization of energy and resources can be achieved by adopting this technology and keeping the devices under surveillance. We can be alerted in case of possible bottlenecks, breakdowns, and damages to the system. Hence, we can save money by using this technology.

10. **Better quality of life**: All the applications of this technology culminate in increased comfort, convenience, and better management, thereby improving the quality of life.

1.5 DISADVANTAGES OF IoT

Some of the disadvantages of IoT are detailed below:

1. **Compatibility**: Currently, there is no international standard of compatibility for the tagging and monitoring equipment. I believe this disadvantage is the most easy to overcome. The manufacturing companies of these equipment just need to agree to a standard, such as Bluetooth, USB, etc. This is nothing new or innovative.

2. **Complexity**: As with all complex systems, there are more opportunities of failure. With the Internet of Things, failures could sky-rocket. For instance, let's say that both you and your spouse each get a message saying that your milk has expired, and both of you stop at a store on your way home, and you both purchase milk. As a result, you and your spouse have purchased twice the amount that you both

need. Or perhaps a bug in the software ends up automatically ordering a new ink cartridge for your printer each and every hour for a few days, or at least after each power failure, when you only need a single replacement.

3. **Privacy/Security:** With all of this IoT data being transmitted, the risk of losing privacy increases. For instance, how well encrypted will the data be kept and transmitted with? Do you want your neighbors or employers to know what medications that you are taking or your financial situation?

4. **Safety:** Imagine if a notorious hacker changes your prescription. Or, alternatively, if a store automatically ships you an equivalent product to which you are allergic, or a flavor that you do not like, or a product that has already expired. As a result, safety is ultimately in the hands of the consumer to verify any and all automation.

   As all the household appliances, industrial machinery, public sector services such as water supply and transport, and many other devices are connected to the Internet, a lot of information is available on it. This information is prone to attack by hackers. It would be disastrous if private and confidential information is accessed by unauthorized intruders.

5. **Compatibility:** As devices from different manufacturers will be interconnected, the issue of compatibility in tagging and monitoring crops up. Although this disadvantage may drop off if all the manufacturers agree to a common standard, even after that, technical issues will persist. Today, we have Bluetooth-enabled devices and compatibility problems exist even in this technology! Compatibility issues may result in people buying appliances from a certain manufacturer, leading to its monopoly in the market.

6. **Complexity:** The IoT is a diverse and complex network. Any failure or bugs in the software or the hardware will have serious consequences. Even power failure can cause a lot of inconvenience.

7. **Lesser Employment of Menial Staff:** Unskilled workers and helpers may end up losing their jobs given the effect of automation of daily activities. This can lead to unemployment issues in the society. This is a problem with the advent of any technology and can be overcome with education. With daily activities becoming automated, naturally, there will be fewer human resources will be required; this will primarily impact unskilled workers and less educated staff. This may create a wider unemployment issue in society.

8. **Technology Takes Control of Life:** Our lives will be increasingly controlled by technology, and we will become increasingly dependent upon it. The younger generation is already becoming addicted to technology for every little thing. We have to decide how much of our daily lives we are willing to mechanize and have controlled by technology.

1.6 DIFFERENT IoT SCENARIOS

Imagine a scenario in which:

- Your fridge can identify that you have run out of milk; it contacts the supermarket and orders the quantity you usually need, and also informs you by sending a message on your phone!
- Your alarm rings at 6:30 a.m.; you wake up and switch it off. As soon as you switch off your alarm, it conveys to the boiler to heat water to a temperature you prefer and also the coffee maker starts brewing coffee!
- You are on your way while returning home from work and you use an app on your mobile to switch on the lights, the AC in your home, and tune the TV to your favorite channel so that your house is ready to welcome you before you even open your door!
- What would really make a refrigerator "smart" would be if it could read tags and alert owners when their food is about to reach their expiry date, for example. Alternatively, perhaps, it could refer to an online calendar and make orders on a regular basis for certain items to be delivered.

1.7 IoT CHARACTERISTICS

1. **Connectivity**: This needs little further explanation. With everything going on in IoT devices and hardware, with sensors and other electronics and connected hardware and control systems there needs to be a connection between various levels.
2. **Things**: Anything that can be tagged or connected as such as it's designed to be connected. From sensors and household appliances to tagged livestock. Devices can contain sensors or sensing materials can be attached to devices and items.
3. **Data**: Data is the glue of the Internet of Things, the first step towards action and intelligence.
4. **Communication**: Devices get connected so they can communicate data and this data can be analyzed. Communication can occur over short distances or over a long range to very long range. Examples: Wi-Fi, LPWA network technologies such as LoRa or NB-IoT.

   LPWA, at its core, is a public wide-area *network* (WAN) with the capability to connect tens of billions of devices spread out over hundreds of millions of square miles. *LPWA* delivers lower data rates than a traditional WAN (like an LTE *network*).
5. **Intelligence**: The aspect of intelligence as in the sensing capabilities in IoT devices and the intelligence gathered from Big Data analytics (also Artificial Intelligence).

6. **Action**: The consequence of intelligence. This can be manual action, action based upon debates regarding phenomena (for instance in smart factory decisions) and automation, often the most important piece.
7. **Ecosystem**: The place of the Internet of Things from a perspective of other technologies, communities, goals and the picture in which the Internet of Things fits. The Internet of Everything dimension, the platform dimension and the need for solid partnerships.

1.8 APPLICATIONS OF IoT

IoT solutions are widely used in numerous companies across industries. Some of the most common IoT applications are given in Table 1.2.

Table 1.2 Some common IoT applications

Application type	Description
Smart thermostats	Helps you to save resource on heating bills by knowing your usage patterns.
Connected cars	IoT helps automobile companies handle billing, parking, insurance, and other related stuff automatically.
Activity trackers	Helps you to capture heart rate pattern, calorie expenditure, activity levels, and skin temperature on your wrist.
Smart outlets	Remotely turn any device on or off. It also allows you to track a device's energy level and get custom notifications directly into your smartphone.
Parking sensors	IoT technology helps users to identify the real-time availability of parking spaces on their phone.
Connect health	The concept of a connected health care system facilitates real-time health monitoring and patient care. It helps in improved medical decision-making based on patient data.
Smart city	Smart city offers all types of use cases which include traffic management to water distribution, waste management, etc.
Smart home	Smart home encapsulates the connectivity inside your homes. It includes smoke detectors, home appliances, light bulbs, windows, door locks, etc.
Smart supply chain	Helps you in real time tracking of goods while they are on the road, or getting suppliers to exchange inventory information.

1.9 CHALLENGES OF THE INTERNET OF THINGS (IoT)

At present, the IoT faces many challenges, such as:

- Insufficient testing and updating
- Concern regarding data security and privacy

- Software complexity
- Data volumes and interpretation
- Integration with AI and automation
- Devices require a constant power supply which is difficult
- Interaction and short-range communication

Key benefits of IoT technology are as follows:

- **Technical Optimization:** IoT technology helps a lot in improving technologies and making them better. Example, with IoT, a manufacturer is able to collect data from various car sensors. The manufacturer analyzes them to improve its design and make them more efficient.
- **Improved Data Collection:** Traditional data collection has its limitations and its design for passive use. IoT facilitates immediate action on data.
- **Reduced Waste:** IoT offers real-time information leading to effective decision-making and the management of resources. For example, if a manufacturer finds an issue in multiple car engines, he can track the manufacturing plan of those engines and solves this issue with the manufacturing belt.
- **Improved Customer Engagement:** IoT allows you to improve customer experience by detecting problems and improving the process.

Chapter 2

IoT architecture

The term "Internet of Things" (IoT) refers to the process by which devices or number of objects is connected to one another over the Internet. These items could include a heart rate monitor, a smart remote or a smart car.

Communication devices in IoT:

Sensors: Devices that transforms physical factors into electrical impulses, such as temperature and motion. For instance, automated farm monitoring can display the present state of the crops, such as how many require water and how much water will be applied to satisfy those needs. This is all due to the Internet of Things (IoT): first, the temperature sensor attached to the plant pot detects the low temperature, then it uses any of the microprocessor platforms such as Raspberry Pi, Arduino boards, etc., then it uses Internet options like Wi-Fi and Bluetooth to receive sensor signals; finally, it notifies the person and the motion sensor attached to the tap, which activates to pour it.

Actuators: These are the devices that simply translate electrical signals into physical movements, acting as the opposite of sensors.

RFID Tags: Wireless microchips are used to tag everything over them for automatic, one-of-a-kind identification. These can be found on credit cards, car ignition keys, and other items. IoT's primary objective is the connectivity of things; hence RFID tags and IoT technology work hand in hand and are utilized to offer a unique identifier for the linked "things" in IoT.

2.1 PHYSICAL DESIGN OF IoT

2.1.1 Things in IoT

The "things" in IoT refers to IoT devices which have unique identities and can perform remote sensing, actuating, and monitoring capabilities. IoT devices can exchange data with other connected devices and applications.

DOI: 10.1201/9781003307488-2

The design of an IoT system corresponds to the number of individual node devices with their associated protocols which are being utilized for the creation of a functional IoT ecosystem. In this system, each and every node device can be used to perform tasks like remote sensing, actuating, monitoring, and so on by relying on physically connected devices. It may also be capable of transmitting information through different types of wireless or wired connections (Figure 2.1).

Connectivity: Devices such as USB hosts and the Ethernet are used for connectivity between the devices and the server.

Processor: A processor like a CPU and other units are used to process the data. These data are further used to improve the decision quality of an IoT system.

Audio/Video Interfaces: An interface such as HDMI and RCA devices is used to record audio and videos in a system.

Input/Output Interface: To give input and output signals to sensors, and actuators we use things like UART, SPI, CAN, etc.

Storage Interfaces: Things like SD, MMC, and SDIO are used to store the data generated from an IoT device. Other things, like DDR and GPU, are used to control the activity of an IoT system.

The devices generate data, and the data is used to perform analysis and carry out operations to improve the system. For example, a moisture sensor is used to obtain the moisture data from a location, and the system analyses it to give an output. Figure 2.2 shows different types of IoT devices.

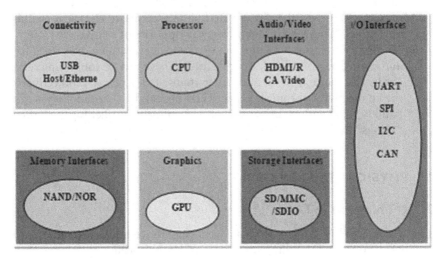

Figure 2.1 Things in IoT.

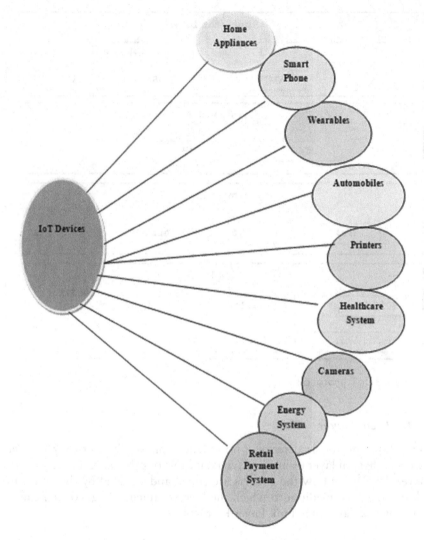

Figure 2.2 IoT devices.

2.1.2 IoT Protocols

These protocols are used to establish communication between a node device and a server over the Internet. This can be used to send commands to an IoT device and to receive data from it. We use different types of protocols that are present on both the server side and the client side and these protocols are managed by network layers like application, transport, network, and link layer, as shown in Figure 2.3.

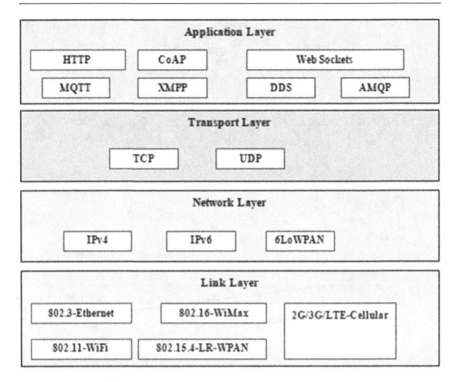

Figure 2.3 IoT protocols.

2.1.2.1 Link layer

Link layer protocols determine how data is physically sent over the network's physical layer or medium (coaxial cable or other or radio wave). This layer determines how the packets are coded and signaled by the hardware device over the medium to which the host is attached (e.g. coaxial cable). Here we explain some Link Layer Protocols:

- **802.3-Ethernet**: Ethernet is a set of technologies and protocols that are used primarily in LANs. This was first standardized in the 1980s by the IEEE 802.3 standard. IEEE 802.3 defines the physical layer and the medium access control (MAC) sub-layer of the data link layer for wired Ethernet networks. Ethernet is classified into two categories: classic Ethernet and switched Ethernet. These standards provide data rates from10 Mb/s to 40Gb/s and higher.
- **802.11-WiFi**: IEEE 802.11 is part of the IEEE 802 set of LAN protocols and specifies the set of media access control (MAC) and physical layer (PHY) protocols for implementing wireless local area network (WLAN) Wi-Fi computer communication in various frequencies, including but not limited to 2.4 GHz, 5 GHz, and 60 GHz frequency bands. These standards provide data rates from1 Mb/s up to 6.75 Gb/s.

- **802.16-Wi-MAX**: The standard for Wi-MAX technology is a standard for Wireless Metropolitan Area Networks (WMANs) that has been developed by working group number 16 of IEEE 802, specializing in point-to-multipoint broadband wireless access. These standards provide data rates from 1.5 Mb/s to 1 Gb/s.
- **802.15.4-LR-WPAN**: A collection of standards for low-rate wireless personal area network. The IEEE's 802.15.4 standard defines the MAC and PHY layer used by, but not limited to, networking specifications such as Zigbee, 6LoWPAN, Thread and WiSUN. The standards provide low-cost and low-speed communication for power-constrained devices. These standards provide data rates from 40 Kb/s to 250 Kb/s.
- **2G/3G/4G-Mobile Communication**: These are different types of telecommunication generations. IoT devices based on these standards can communicate over the cellular networks. Data rates for these standards range from 9.6 Kb/s up to 100 Mb/s.

2.1.2.2 Network layer

The network layer is responsible for the sending of IP data grams from the source network to the destination network. It performs the host addressing and packet routing. We used IPv4 and IPv6 for host identification. IPv4 and IPv6 are hierarchical IP addressing schemes.

IPv4: **Internet Protocol address (IP address)** is a numerical label assigned to each device connected to a computer network that uses the Internet Protocol (IP) for communication. An IP address serves two main functions: host or network interface identification and location addressing. Internet Protocol version 4 (IPv4) defines an IP address as a 32-bit number. However, because of the growth of the Internet and the depletion of available IPv4 addresses, a new version of IP (IPv6), using 128 bits for the IP address, was standardized in 1998. IPv6 deployment has been ongoing since the mid-2000s.

IPv6: **Internet Protocol version 6 (IPv6)** is the successor to IPv4. IPv6 was developed by the Internet Engineering Task Force (IETF) to deal with the long-anticipated problem of IPv4 address exhaustion. In December 1998, IPv6 became a Draft Standard for the IETF, who subsequently ratified it as an Internet Standard on 14 July 2017. IPv6 uses a 128-bit address, theoretically allowing 2^{128}, or approximately 3.4 × 10^{38} addresses.

6LoWPAN: This is an acronym forIPv6 over Low-Power Wireless Personal Area Networks. 6LoWPAN is the name of a concluded working group in the Internet area of the IETF. This protocol allows for the smallest devices with limited processing ability to transmit information wirelessly using an Internet protocol. 6LoWPAN can communicate with 802.15.4 devices as well as other types of devices on an IP network link like WiFi.

2.1.2.3 Transport layer

This layer provides functions such as error control, segmentation, flow control, and congestion control. Thus, this layer protocols provide end-to-end message transfer capability independent of the underlying network:

TCP: TCP (Transmission Control Protocol) is a standard that defines how to establish and maintain a network conversation through which application programs can exchange data. TCP works with the IP, which defines how computers send packets of data to each other. Together, TCP and IP are the basic rules defining the Internet. The Internet Engineering Task Force (IETF) defines TCP in the Request for Comment (RFC) standards document number 793.

UDP: User Datagram Protocol (UDP) is a Transport Layer protocol. UDP is a part of Internet Protocol suite, referred as the UDP/IP suite. Unlike TCP, it is an unreliable and connectionless protocol. Thus, there is no need to establish connection prior to data transfer. It is useful for time-sensitive applications that have very small data units to exchange and do not want the overhead of connection setup.

2.1.2.4 Application layer

Application layer protocols define how the applications interface with the lower-layer protocols to send over the network. Port numbers are used for application addressing (e.g. port 80 for HTTP, port 22 for SSH etc.).

HTTP: *Hypertext Transfer Protocol (HTTP)* is an application-layer protocol for transmitting hypermedia documents, such as HTML. It was designed for communication between web browsers and web servers, but it can also be used for other purposes. HTTP follows a classical client–server model, with a client opening a connection to make a request, then waiting until it receives a response. HTTP is a stateless protocol, meaning that the server does not keep any data (state) between two requests. HTTP includes commands such as GET, PUT, POST, HEAD, TRACE, OPTIONs etc. HTTP uses URI's (Uniform Resource Identifiers) to identify HTTP resources.

CoAP: CoAP-Constrained Application Protocol is a specialized Internet Application Protocol for constrained devices, as defined in RFC 7252. It enables devices to communicate over the Internet. The protocol is especially targeted at constrained hardware such as 8-bit microcontrollers, low-power sensors, and similar devices which cannot run on HTTP or TLS. It is a request response model; however, it runs on top of UDP instead of TCP. CoAP uses client–server architecture where clients communicate with servers using connectionless datagrams.

WebSocket: The WebSocket Protocol enables two-way communication between a client running untrusted code in a controlled environment and a remote host that has opted in to communications from that code. The security model used for this is the origin-based security model commonly used by web browsers.

MQTT: MQTT is a machine-to-machine (M2M)/Internet of Things connectivity protocol. It was designed as an extremely lightweight publish/subscribe messaging transport and useful for connections with remote locations where a small code footprint is required and/or network bandwidth is at a premium. *It is well suited for constrained environments where the devices have limited processing and memory resources and network bandwidth is low. MQTT uses a client server architecture where the client (IoT device) connects to the server (MQTT broker) and publishes messages to topics on the server. The broker forwards the messages to the clients subscribed to topics.*

XMPP: Extensible Messaging and Presence Protocol (XMPP) is a communication protocol for message-oriented middleware based on XML (Extensible Markup Language). It enables the near-real-time exchange of structured, yet extensible data between any two or more network entities.

DDS: The Data Distribution Service (DDS) is a middleware protocol and API standard for data-centric connectivity from the Object Management Group (OMG). It integrates the components of a system together, providing low-latency data connectivity, extreme reliability, and a scalable architecture that business and mission-critical Internet of Things (IoT) applications need.

AMQP: The AMQP IoT protocols consist of a hard and fast of components that route and save messages within a broker carrier, with a set of policies for wiring the components together. The AMQP protocol enables patron programs to talk to the dealer and engage with the AMQP model. It supports both point to point and publisher/subscriber models, routing and queueing.

2.2 LOGICAL DESIGN OF IoT

This design of the IoT system refers to an abstract representation of the entities and processes without going into the low-level specifics of the implementation. To understand the logical design of IoT, some terms are used:

- IoT functional blocks
- IoT communication models
- IoT communication APIs

2.2.1 IoT functional blocks

An IoT system contains a number of functional blocks, which provides the system the capabilities for identification, sensing, actuation, communication as well as management. There are number of functional blocks including:

- **Device**: an IoT system comprises devices that provide sensing, actuation, monitoring, and control functions.
- **Communication**: handles the communication for the IoT system.
- **Services**: services for device monitoring, device control service, data publishing services and services for device discovery.
- **Management**: this block provides various functions to govern the IoT system.
- **Security**: this block secures the IoT system and by providing functions such as authentication, authorization, message integrity, and data security.
- **Application**: an interface that the users can use to control and monitor various aspects of the IoT system. Application also allows users to view the system status and view or analyze the processed data (Figure 2.4).

2.2.2 IoT communication models

- **Request–Response Model**: This is actually a communication model based on a client–server concept in which the client sends the request to the server and the server responds to that request. Whenever the server receives a request from the client, it decides how to respond. This means firstly that it will fetch the data, retrieve the resource representation, prepare the response and then send the response to the client. Request–response is a stateless communication model and each request–response pair is independent of others.

  HTTP works as a request–response protocol between a client and server. A web browser may be the client and an application on a computer that hosts a web site may be the server.

  Example: A client (browser) submits an HTTP request to the server; then the server returns a response to the client. The response

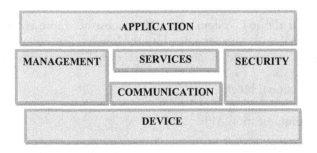

Figure 2.4 Logical design of IoT.

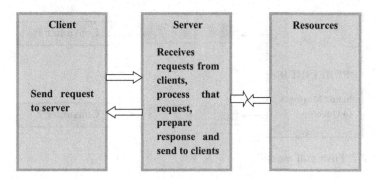

Figure 2.5 Request response communication model.

contains status information about the request and may also contain the requested content as shown in Figure 2.5.

- **Publish–Subscribe Model**: This is a communication model that involves publishers, brokers, and consumers. The publishers are the source of data. They will send the data to the topics which are managed by the broker. Publishers are unaware of the consumers. Consumers subscribe to the topics which are managed by the broker. When the broker receives the data for a topic from the publisher, it sends the data to all the subscribed consumers as shown in Figure 2.6.
- **Push–Pull Model**: It is a communication model in which the data producers push the data to queues and the consumers pull the data from the queues. Producers do not need to be aware of the consumers. Queues help in decoupling the messaging between the producers and consumers. Queues also act as a buffer which helps in situations when there is a mismatch between the rate at which the producers push data and the rate at which the consumer pull data as shown in Figure 2.7.
- **Exclusive Pair Model**: This is a bidirectional, fully duplex communication model that uses a persistent connection between the client and server. Firstly, the connection is setup; it will remain open until the

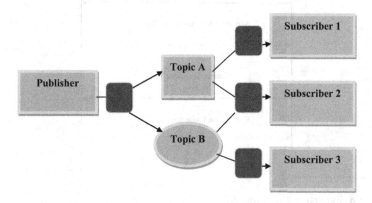

Figure 2.6 Publish-subscribe model.

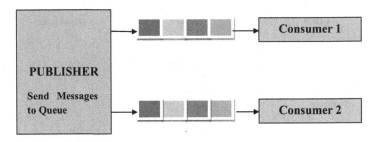

Figure 2.7 Push pull model.

client sends a request to close the connection. The client and the server can send messages to each other after connection setup. Exclusive pair is state full communication model and the server is aware of all the open connections. Figure 2.8 shows the client–server interactions in the exclusive pair model.

2.2.3 IoT communication APIs

There are generally two APIs for IoT Communication. These IoT Communication APIs are:

- REST-based Communication APIs
- WebSocket-based Communication APIs

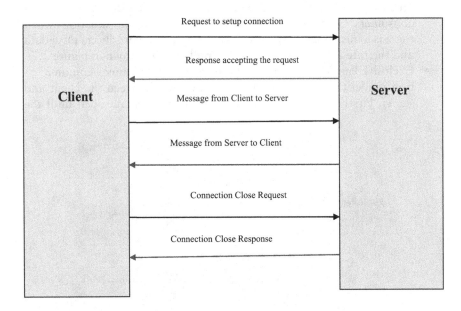

Figure 2.8 Exclusive pair model.

2.2.3.1 REST-based communication APIs

Representational state transfer (REST) is a set of architectural principles by which you can design Web services. The Web APIs that focus on systems' resources and how resource states are addressed and transferred. REST APIs that follow the request response communication model, the rest architectural constraint apply to the components, connector, and data elements within a distributed hypermedia system. The rest architectural constraints are as follows:

Client-server: The principle behind the client-server constraint is the separation of concerns. For example, clients should not be concerned with the storage of data which is the concern of the server. Similarly, the server should not be concerned about the user interface, which is the concern of the client. Separation allows client and server to be independently developed and updated.

Stateless: Each request from client to server must contain all the information necessary to understand the request and cannot take advantage of any stored context on the server. The session state is kept entirely on the client.

Cache-able: Cache constraints requires that the data within a response to a request be implicitly or explicitly leveled as cache-able or non-cache-able. If a response is cache-able, then a client cache is given the right to reuse that response data for later, equivalent requests. Caching can either partially or completely eliminate some instructions and improve efficiency and scalability.

Layered system: This constrains the behavior of components such that each component cannot see beyond the immediate layer with which they are interacting. For example, the client cannot tell whether it is connected directly to the end server or to an intermediately along the way. System scalability can be improved by allowing intermediaries to respond to requests instead of the end server, without the client having to do anything different.

Uniform interface: This constraint requires that the method of communication between client and server must be uniform. Resources are identified in the requests (by URLs in web-based systems) and are themselves separate from the representations of the resources data returned to the client. When a client holds a representation of resources, it has all the information required to update or delete the resource you (provided the client has required permissions). Each message includes enough information to describe how to process the message.

Code on demand: Servers can provide executable code or scripts for clients to execute in their context. This constraint is the only one that is optional.

A RESTful web service is "Web API" implemented using HTTP and REST principles. REST is most popular IoT Communication APIs. Table 2.1 shows

Table 2.1 Request methods and actions

Uniform Resource Identifier (URI)	GET	PUT	PATCH	POST	DELETE
Collection, such as https://api.example.com/resources/	List the URIs and perhaps other details of the collection's members.	Replace the entire collection with another collection.	Not generally used	Create a new entry in the collection. The new entry's URI is assigned automatically and is usually returned by the operation.	Delete the entire collection.
Element, such as https://api.example.com/resources/item5	Retrieve a representation of the addressed member of the collection, expressed in an appropriate Internet media type.	Replace the addressed member of the collection, or if it does not exist, create it.	Update the addressed member of the collection.	Not generally used. Treat the addressed member as a collection in its own right and create a new entry within it.	Delete the addressed member of the collection.

how the client send requests to URI are using the methods defined by the HTTP protocol.

2.2.3.2 Web Socket-based communication API

Web Socket APIs allows bidirectional, full duplex communication between clients and servers, as shown in Figure 2.8. It follows the exclusive pair communication model. Unlike request–response models such as REST, the Web Socket APIs allow full duplex communication and do not require a new connection to be setup for each message to be sent.

Web Socket communication begins with a connection setup request sent by the client to the server. The request (called a Web Socket handshake) is sent over HTTP and the server interprets it is an upgrade request. If the server supports a Web Socket protocol, the server responds to the web socket handshake response. After the connection setup, the client and server can send data/messages to each other in full duplex mode. Web Socket API reduces the network traffic and latency as there is no overhead for connection setup and termination requests for each message. Web Socket is suitable for IoT applications that have low latency or high throughput requirements. Thus, Web Socket is the most suitable IoT Communication API for an IoT system (Figure 2.9).

1. Request to setup Web Socket Connection
2. Response accepting the request

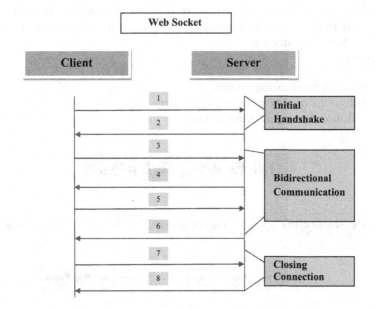

Figure 2.9 Web Socket API's.

3. Data Frame
4. Data Frame
5. Data Frame
6. Data Frame
7. Connection Close Request
8. Connection Close Response

2.3 IoT-ENABLING TECHNIQUES

IoT is enabled by several technologies, including wireless sensor networks, cloud computing, Big Data analytics, communication protocols, embedded systems etc., as shown in Figure 2.10. This section describes each and every enabling techniques used in IoT:

- Wireless sensor network
- Cloud computing
- Big Data analytics
- Communications protocols
- Embedded system

2.3.1 Wireless Sensor Network (WSN)

This comprises a number of distributed devices with sensors which are used to monitor the environmental and physical conditions. A wireless sensor network consists of end nodes, routers, and coordinators. End nodes have several sensors attached to them where the data is passed to a coordinator with the help of routers. The coordinator also acts as the gateway that connects WSN to the Internet.

Example:

- Weather monitoring system
- Indoor air quality monitoring system
- Soil moisture monitoring system

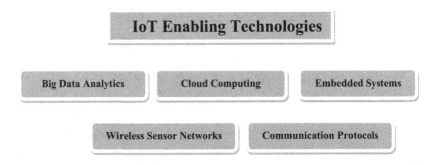

Figure 2.10 IoT enabling techniques.

- Surveillance system
- Health monitoring system

2.3.2 Cloud computing

Cloud computing provides us with the means by which we can access applications as utilities over the Internet. Within this context, the notion of the cloud means something which is present in remote locations. Using cloud computing, users can access any resources from anywhere like databases, web servers, storage, any device, and any software over the Internet.

Characteristics

- Broad network access
- On-demand self-services
- Rapid scalability
- Measured service
- Pay-per-use

It provides different services like:

- **IaaS (Infrastructure as a Service)**: IaaS (Infrastructure as a Service) is also known as **Hardware as a Service (HaaS)**. It is one of the layers of the cloud computing platform and allows customers to outsource their IT infrastructures, such as servers, networking, processing, storage, virtual machines, and other resources. In traditional hosting services, IT infrastructure was rented out for a specific period of time, with pre-determined hardware configuration. The client paid for the configuration and time, regardless of the actual use. With the help of the IaaS cloud computing platform layer, clients can dynamically scale the configuration to meet changing requirements and are billed only for the services actually used.

  Major IaaS providers include Google Compute Engine, Amazon Web Services and Microsoft Azure, among others.

  Examples: Web Hosting, Virtual Machine etc. (Figure 2.11).
- **PaaS (Platform as a Service)**: Platform as a Service (PaaS) provides a runtime environment. It allows programmers to easily create, test, run, and deploy web applications. One can purchase these applications from a cloud service provider on a pay-per-use basis and access them using the Internet connection. In PaaS, back-end scalability is managed by the cloud service provider, so end-users do not need to worry about managing the infrastructure. PaaS includes infrastructure (servers, storage, and networking) and platform (middleware, development tools, database management systems, business intelligence, and more) to support the web application life cycle.

  Example: Google App Engine, Force.com, Joyent, Azure (Figure 2.12).

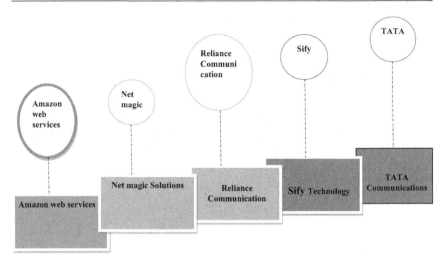

Figure 2.11 IaaS providers.

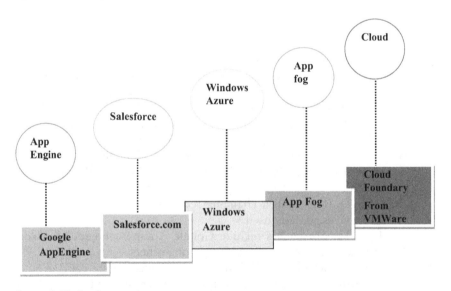

Figure 2.12 PaaS providers.

- **SaaS (Software as a Service)**: This is also known as "On-Demand Software". It is a software distribution model in which services are hosted by a cloud service provider. These services are available to end-users over the Internet, meaning that the end-users do not need to install any software on their devices to access these services. SaaS applications are otherwise known as web-based software, on-demand software, or hosted software. SaaS applications run on a SaaS provider's service and they manage security availability and performance.

  Example: Google Docs, Gmail, office etc. (Figure 2.13).

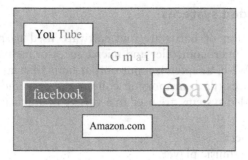

Figure 2.13 SaaS providers.

2.3.3 Big Data analytics

This refers to the method of studying massive volumes of data, or the so-called Big Data. It involves the collection of data whose volume, velocity, or variety is simply too massive and tough to store, control, process, and examine the data using traditional databases. Big Data is gathered from a variety of sources, including social network videos, digital images, sensors, and sales transaction records.

There are several steps involved in analyzing Big Data:

- Data cleaning
- Munging (Wrangling)
- Processing
- Visualization

Examples of Big Date might include:

- Bank transactions
- Data generated by IoT systems for the location and tracking of vehicles
- E-commerce and in Big-Basket
- Health and fitness data generated by a IoT system such as through a fitness band

2.3.4 Communications protocols

They are the backbone of IoT systems and enable network connectivity and linking to applications. Communication protocols allow devices to exchange data over the network. Multiple protocols often describe different aspects of a single communication. A group of protocols designed to work together is known as a protocol suite; when implemented in software they are a protocol stack.

They are used in:

- Data encoding
- Addressing schemes

2.3.5 Embedded systems

It is a combination of hardware and software used to perform special tasks. It includes microcontroller and microprocessor memory, networking units (Ethernet Wi-Fi adapters), input output units (display keyword etc.), and storage devices (flash memory).It collects the data and sends it to the Internet. Embedded systems are used in a variety of applications:

1. Digital camera
2. DVD player, music player
3. Industrial robots
4. Wireless routers etc.

2.4 IoT LEVELS

An IoT architecture element varies on the basis of applications used. On the basis of this fact, various levels are defined for an IoT system. Let us understand these IoT levels with their elements and examples of their usage. Developing an **IoT Level Template** system consists of the following components:

1. **Device:** These may be sensors or actuators capable of identifying, remote sensing, or monitoring.
2. **Resources:** These are software components on IoT devices for accessing and processing. Storing software components or controlling actuators connected to the device. Resources also include software components that enable network access.
3. **Controller Service:** This is a service that runs on the device and interacts with web services. The controller service sends data from the device to the web service and receives commands from the application via web services for controlling the device.
4. **Database:** Stores data generated from the device.
5. **Web Service:** This provides a link between IoT devices, applications, databases, and analysis components.
6. **Analysis Component:** It performs an analysis of the data generated by the IoT device and generates results in a form which are easy for the user to understand.
7. **Application:** It provides a system for the user to view the system status and view product data. It also allows users to control and monitor various aspects of the IoT system.

Let us take the example of an air conditioner whose temperature has to be monitored to understand IoT levels.

2.4.1 IoT level 1

This level consists of air conditioner, temperature sensor, data collection and analysis, and control & monitoring apps. The data sensed is stored locally. The data analysis is done locally. Monitoring and control is carried out using a mobile app or a web app. The data generated in this level application is not huge. All of the control actions are performed through the Internet.

- Example: Room temperature is monitored using a temperature sensor and data is stored/analysed locally. Based on the analysis, a control action is triggered using a mobile app or it can just help in status monitoring.

  Example: We can understand with the help of an example. Let's look at the IoT device that monitors the lights in a house. The lights are controlled through switches. The database has maintained the status of each light; in addition, REST services deployed locally allow retrieving and updating the state of each light and trigger the switches accordingly. To control the lights and applications, the application has an interface. The device is connected to the Internet and hence the application can also be accessed remotely (Figure 2.14).

2.4.2 IoT level 2

This level consists of the air conditioner, the temperature sensor, and Big Data (Bigger than level-1, data analysis done here), cloud and the control & monitoring app. This level-2 is more complex than level-1. In addition, the rate of sensing is faster than at level-1. This level has a voluminous size

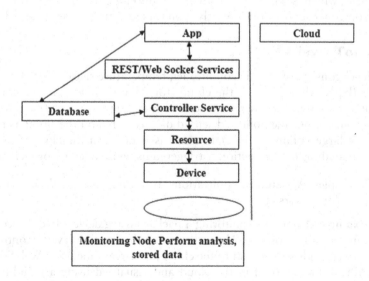

Figure 2.14 IoT level 1.

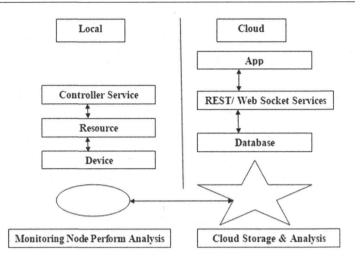

Figure 2.15 IoT level 2.

of data, and hence cloud storage is used. Data analysis is carried out locally. The cloud is used for only storage purposes. Based on data analysis, control action is triggered using the web app or the mobile app.

- Examples: Agriculture applications, room freshening solutions based on odor sensors etc.
  Example: A cloud-based application is used for monitoring and controlling the IoT system. A single node monitors the soil moisture in the field, which is sent to the database on the cloud using REST APIS. The controller service continuously monitors moisture levels (Figure 2.15).

2.4.3 IoT level 3

This level consists of the air conditioner, temperature sensor, Big Data collection (bigger than level-1), the cloud (for data analysis) and the control & monitoring app. Data here is voluminous i.e. Big Data. The frequency of data sensing is fast and collected sensed data is stored on the cloud as there is such a large volume of it. Data analysis is done on the cloud side and based on analysis control action is triggered using mobile app or web app.

- Examples: Agriculture applications, room freshening solutions based on odor sensors etc.

  Example: A node is monitoring a package using devices like an accelerometer and gyroscope. These devices track vibration levels. Controller service sends sensor data to the cloud in the rear time using Web Socket APL. Data is stored in the cloud and visualized using a cloud-based application. The analysis component triggers an alert if vibration levels cross a threshold (Figure 2.16).

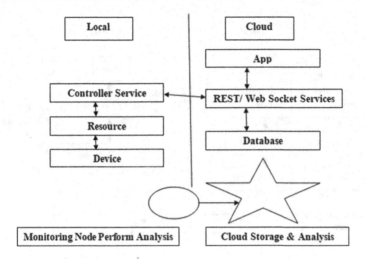

Figure 2.16 IoT level 3.

2.4.4 IoT level 4

This level consists of multiple sensors, data collection and analysis, and a control & monitoring app. At this level-4, multiple sensors are used which are independent of the others. The data collected using these sensors are uploaded to the cloud separately. Cloud storage is used at this level because of the huge data storage required. The data analysis is performed on the cloud and based on which control action is triggered, using either a web app or a mobile app.

> **Example:** Analysis is done on the cloud and the entire IoT system has monitored the cloud using an application. Noise monitoring of an area requires various nodes to function independently of each other. Each has its own controller service. Data is stored in a cloud database (Figure 2.17).

2.4.5 IoT level 5

This level consists of multiple sensors, coordinator node, data collection and analysis and control & monitoring app. This level is similar to level-4, which also has huge data and hence they are sensed using multiple sensors at much faster rate and simultaneously. The data collection and data analysis is performed at the cloud level. Based on analysis, control action is performed using mobile app or web app.

> **Example:** A monitoring system has various components: end nodes collect various data from the environment and send it to the coordinator node. The coordinator node acts as a gateway and allows the data to be transferred to cloud storage using REST API. The controller service on the coordinator node sends data to the cloud (Figure 2.18).

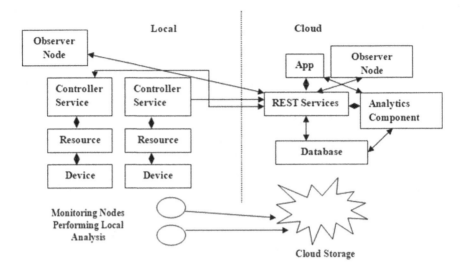

Figure 2.17 IoT level 4.

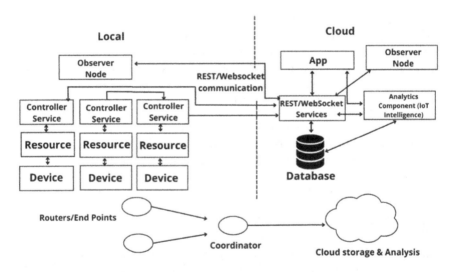

Figure 2.18 IoT level 5.

2.4.6 IoT level 6

At this level, the application is also cloud-based and data is stored in the cloud. Multiple independent end nodes perform sensing and actuation and send data to the cloud. The analytics components analyze the data and store the results in the cloud database. The results are visualized with a

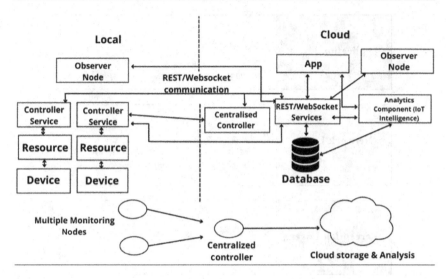

Figure 2.19 IoT level 6.

cloud-based application. The centralized controller is aware of the status of all the end nodes and sends control commands to the nodes.

> **Example**: Weather monitoring consists of sensors that monitor different aspects of the system. The end nodes send data to cloud storage. Analysis of components, applications, and storage areas in the cloud. The centralized controller controls all nodes and provides inputs (Figure 2.19).

2.5 ARCHITECTURAL OVERVIEW OF THE INTERNET OF THINGS (IoT)

Can you imagine a huge variety of smart devices under the centralized control of one "brain"? To a certain extent, it's possible with the evolution of the Internet of Things – the network of physical objects with sensors and actuators, software, and network connectivity that enable these objects to gather and transmit data and fulfill users' tasks.

The effectiveness and applicability of such a system directly correlate with the quality of its building blocks and the way in which they interact, and there are various approaches to IoT architecture. In this, there is hands-on experience of a scalable and flexible IoT architecture. IoT technology has a wide variety of applications and the use of the Internet of Things is growing much faster. Depending upon the different application areas of the Internet of Things, it works accordingly to the way in which it has been designed/developed. But there is no single standard architecture of working which is

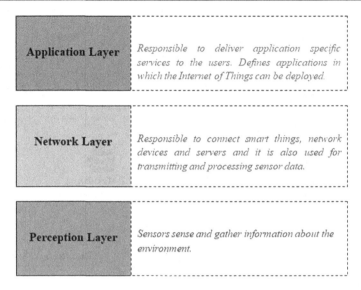

Figure 2.20 Basic architecture of IoT.

Business Layer	*Manages the whole IoT system, including applications, business and profit models.*
Application Layer	*Responsible for delivering application specific services to the user.*
Processing Layer	*Stores, analyses and processes huge amount of data.*
Transport Layer	*Transfers the sensor data between layer through networks as LAN, Bluetooth, RFID etc.*
Perception Layer	*Sensors sense and gather information about the environment*

Figure 2.21 Five layered architecture of IoT.

followed universally. The architecture of IoT depends upon its functionality and implementation in different sectors. Still, there is a basic process flow based upon which IoT is built. The basic fundamental architecture of IoT is i.e., 4 Stage IoT architecture (Figures 2.20–2.22).

From Figure 2.22, it is clear that there are four layers present which can be divided as follows: sensing layer, network layer, data processing layer, and application layer.

These are explained below.

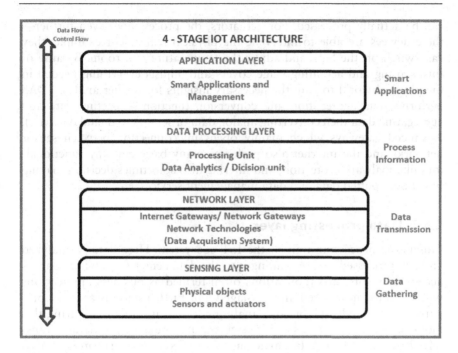

Figure 2.22 Four layered architecture of IoT.

2.5.1 Sensing layer

Sensors, actuators, and devices are present in this sensing layer. The outstanding feature about sensors is their ability to convert the information obtained in the outer world into data for analysis. In other words, it's important to start with the inclusion of sensors in the four stages of an IoT architecture framework to get information in an appearance that can be actually processed. For actuators, the process goes even further—these devices are able to intervene the physical reality. For example, they can switch off the light and adjust the temperature in a room. Because of this, sensing and actuating stage covers and adjusts everything needed in the physical world to gain the necessary insights for further analysis. These sensors or actuators accepts data (physical/environmental parameters), processes data, and emits data over network.

2.5.2 Network layer

Internet/network gateways, Data Acquisition System (DAS) are present in this layer. The outstanding feature about sensors is their ability to convert the information obtained in the outer world into data for analysis. In other words, it's important to start with the inclusion of sensors in the four stages of an IoT architecture framework to get information in an appearance that can be actually processed. For actuators, the process goes even further—these devices are able to intervene the physical reality. For example, they can switch off the light and adjust the temperature in a room. Because of this, sensing and actuating stage covers and adjusts everything needed in the physical world to gain the necessary insights for further analysis. DAS performs data aggregation and conversion function (collecting data and aggregating data then converting analog data of sensors to digital data etc.). Advanced gateways which mainly opens up connection between sensor networks and the Internet also performs many basic gateway functionalities like malware protection, and filtering also sometimes decision-making based on inputted data and data management services, etc.

2.5.3 Data processing layer

This is the processing unit of the IoT ecosystem. Here data is analyzed and pre-processed before sending it to the data center from where data is accessed by software applications often termed as business applications where data is monitored and managed and further actions are also prepared. So here edge IT or edge analytics comes into picture. During this moment among the stages of IoT architecture, the prepared data is transferred to the IT world. In particular, edge IT systems perform enhanced analytics and pre-processing here. For example, it refers to machine learning and visualization technologies. At the same time, some additional processing may happen here, prior to the stage of entering the data center. Likewise, Stage 3 is closely linked to the previous phases in the building of architecture of IoT. Because of this, the location of edge IT systems is close to the one in which sensors and actuators are situated, creating a wiring closet. At the same time, the residing in remote offices is also possible.

2.5.4 Application layer

This is the fourth and last layer of IoT architecture. The data center or the cloud is the management stage of data, where data is managed and is used by end-user applications in sectors such as agriculture, healthcare, aerospace, farming, and defense, etc. The main processes on the final stage of IoT architecture happen in data center or cloud. Precisely, it enables in-depth processing, along with a follow-up revision for feedback. Here, the skills of both IT and OT (operational technology) professionals are needed. In other words, the phase already includes the analytical skills of the highest rank, in both the digital and human worlds. Therefore, the data from other sources may be included here to ensure an in-depth analysis. After meeting all the quality standards and requirements, the information is brought back to the physical world but in a processed and precisely analyzed appearance already.

2.5.5 Stage 5 of IoT architecture

In fact, there is an option to extend the process of building a sustainable IoT architecture by introducing an extra stage in it. This refers to initiating a user's control over the structure, if only your result doesn't include full automation, of course. The main tasks here are visualization and management. After including Stage 5, the system turns into a circle where a user sends commands to sensors/actuators (Stage 1) to perform some actions.

2.5.6 IoT architecture example: intelligent lighting

Let's see how our IoT architecture elements work together through the example of smart yard lighting as a part of a smart home, a bright illustration of how an IoT solution simultaneously contributes to user convenience and energy efficiency. There are various ways a smart lighting system can function, and we'll cover basic options in Figure 2.23.

> **Basic components**: Sensors take data from the environment (for example, daylight, sounds, people's movements). Lamps are equipped with actuators to switch the light on and off. A *data take* stores raw data

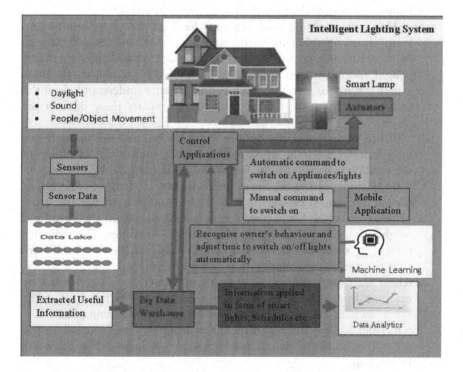

Figure 2.23 Intelligent lighting system.

coming from sensors. A *Big Data warehouse* contains the extracted info smart home dwellers' behavior on various days of the week, energy costs, and more.

Manual monitoring and manual control: Users control a smart lighting system with a *mobile app* featuring the map of the yard. With the app, users can see which lights are on and off and send commands to the control applications that further transmit them to lamp actuators. Such an app can also show which lamps are about to fail.

Data analytics: Analyzing the way users apply smart lighting, their schedules (either provided by users or identified by the smart system), and other info gathered with sensors, data analysts can make and update the algorithms for control applications.

Data analytics also helps in assessing the effectiveness of the IoT system and revealing problems in the way the system works. For example, if a user switches off the light right after a system automatically switches it on and vice versa, there might be gaps in the algorithms, and it's necessary to address them as soon as possible.

Automatic control's options and pitfalls: The sensors monitoring natural light send the data about the light to the cloud. When the daylight is not sufficient (according to the previously stated threshold), the control apps send automatic commands to the actuators to switch on the lamps. The rest of the time the lamps are switched off.

However, a lighting system can be "deceived" by street illumination, lamps from neighboring yards, and any other sources. Extraneous light captured by sensors can make the smart system conclude that it is light enough, and that any lighting should be switched off. Thus, it makes sense to give the smart system a better understanding of the factors that influence lighting and to accumulate these data in the cloud.

When sensors monitor motions and sounds, it's not enough just to switch on the light when movements or sounds are identified in the yard or to switch all the lamps off in the silence. Movements and sounds can be produced, for example, by pets, and cloud applications should distinguish between human voices and movements and those of pets. The same issues apply to noises coming from the street and neighboring houses, and also other sounds. To address this issue, it's possible to store the examples of various sounds in the cloud and compare them with the sounds coming from the sensors.

Machine learning: Intelligent lighting can apply models generated by machine learning, for example, to recognize the patterns of smart home owners' behavior (leaving home at 8 a.m., coming back at 7 p.m.) and accordingly adjust the time when lights are switched on and off (for example, to switch any lamps on 5 minutes before they will be needed).

Analyzing users' behavior in a long-term perspective, a smart system can develop advanced behavior. For example, when sensors don't identify the typical movements and voices of home inhabitants, a smart system can "suppose" that smart home dwellers are on a holiday and adjust its behavior accordingly: for example, the system might occasionally switch on the lights to give the impression that the house is not empty (for security reasons), but also not keep the lights on all the time to reduce energy consumption.

User management options: To ensure efficient user management, the smart lighting system can be designed for several users with role distribution: for example, owner, inhabitants, guests. In this case, the user with the title "*owner*" will have full control over the system (including changing the patterns of smart light behavior and monitoring the status of the yard lamps) and priorities in giving commands (when several users give contradicting commands, the owner's command will be the default instruction); other users will have access to only a limited number of the system's functions. "*Inhabitants*", for example might be enabled to switch the lamps on and off with no opportunity to change settings. Similarly, "*Guests*" will be able to switch lights on and off in some parts of the house and would have no access to controlling the lights, for example, near the garage.

Apart from role distribution, it's essential to consider ownership (as soon as one system can control over 100,000 households, and it's important that a dweller of a smart home manages the lighting in their own yard, and not that of a neighbor).

2.6 REFERENCE MODEL AND ARCHITECTURE

An Architecture Reference Model (ARM) consists of two main parts: a reference model and a reference architecture.

- A reference model describes the domain using a number of sub-models (Figures 2.24 and 2.25).

2.6.1 IoT reference model

The foundation of an IoT Reference Architecture description is an IoT reference model. A reference model describes the domain using a number of sub-models. The domain model of an architecture model captures the main concepts or entities in the domain in question, in this case M2M and IoT.

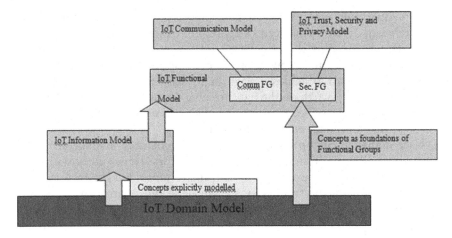

Figure 2.24 IoT model.

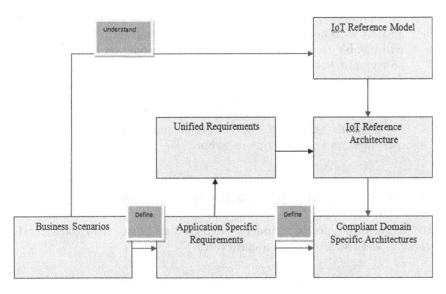

Figure 2.25 IoT reference model.

2.6.1.1 IoT domain model

The domain model captures the basic attributes of the main concepts and the relationship between these concepts. A domain model also serves as a tool for human communication between people working in the domain in question and also between people who work across different domains.

Model notation and semantics (Figure 2.26)

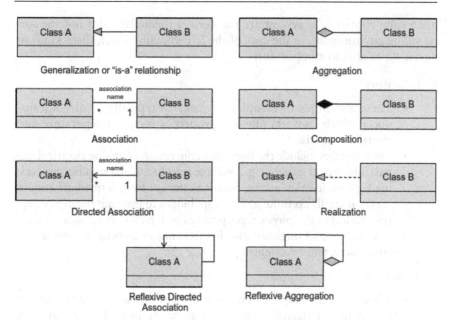

Figure 2.26 IoT domain model.

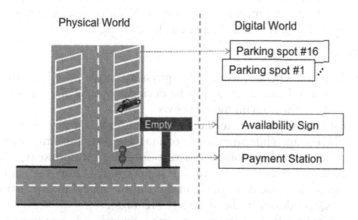

Figure 2.27 Representation of physical world.

Main concepts

The IoT is a support infrastructure for enabling objects and places in the physical world to have a corresponding representation in the digital world (Figure 2.27).

The devices are physical artifacts with which the physical and virtual worlds interact. Devices, as mentioned before, can also be physical entities for certain types of applications, such as management applications when the interesting entities of a system are the devices themselves and not the surrounding environment. In the case of the IoT domain model, three kinds of device types are most important:

1. **Sensors:**

   - These are simple or complex devices that typically involve a transducer which converts physical properties such as temperature into electrical signals.
   - These devices include the necessary conversion of analog electrical signals into digital signals, e.g. a voltage level to a 16-bit number, processing for simple calculations, potential storage for intermediate results, and potentially communication capabilities to transmit the digital representation of the physical property as well receive commands.
   - A video camera might be another example of a complex sensor that could detect and recognize people.

2. **Actuators:**

   - These are also simple or complex devices that involve a transducer that converts electrical signals to a change in a physical property (e.g. turn on a switch or move a motor).
   - These devices also include potential communication capabilities, the storage of intermediate commands, processing, and the conversion of digital signals to analog electrical signals.

3. **Tags:**

   - Tags in general identify the physical entity to which they are attached. In reality, tags can be devices or physical entities but not both, as the domain model shows.
   - An example of a tag as a device is a Radio Frequency Identification (RFID) tag, while an example of a tag as a physical entity might be a paper-printed immutable barcode or a Quick Response (QR) code.
   - Either an electronic device or a paper-printed entity tag contains a unique identification that can be read by optical means (bar codes or QR codes) or radio signals (RFID tags).
   - The reader device operating on a tag is typically a sensor, or sometimes a combined sensor and actuator, as in the case of writable RFID tags.

2.6.1.2 Information model

A virtual entity in the IoT domain model is the "Thing" in the Internet of Things; the IoT information model captures the details of a virtual entity-centric model. Similar to the IoT domain model, the IoT information model is presented using Unified Modeling Language (UML) diagrams (Figure 2.28).

Relationship between core concepts of IoT domain model and IoT information model (Figure 2.29).

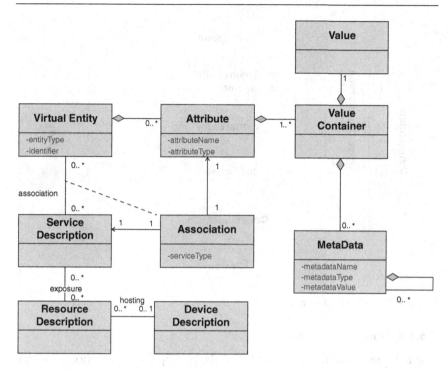

Figure 2.28 IoT information model.

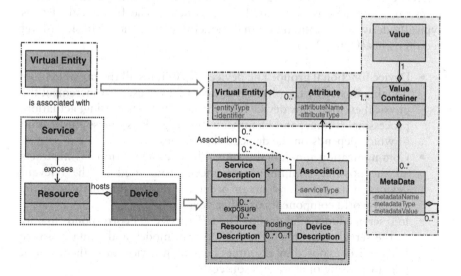

Figure 2.29 IoT domain and information model.

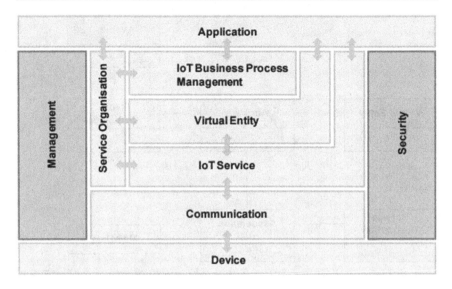

Figure 2.30 IoT function model.

2.6.1.3 Functional model

The IoT functional model aims to describe principally the functional groups (FG) and their interaction with the ARM, while the functional view of a reference architecture describes the functional components of an FG, interfaces, and interactions between the components. The functional view is typically derived from the functional model in conjunction with high-level requirements (Figure 2.30).

- **Device functional group**: The device FG contains all the possible functionality hosted by the physical devices that are used for increment the physical entities. This device functionality includes sensing, actuation, processing, storage, and identification components, the sophistication of which depends on the device's capabilities
- **Communication functional group**: The communication FG abstracts all the possible communication mechanisms used by the relevant devices in an actual system in order to transfer information to the digital world components or other devices.
- **IoT service functional group**: The IoT service FG corresponds mainly to the service class from the IoT domain model, and contains single IoT services exposed by resources hosted on devices or in the network (e.g. processing or storage resources).
- **Virtual entity functional group**: The virtual entity FG corresponds to the virtual entity class in the IoT domain model, and contains the necessary functionality to manage associations between virtual entities with themselves as well as associations between virtual entities and related IoT services, i.e. the association objects for the IoT information

model. Associations between virtual entities can be either static or dynamic, depending on the mobility of the physical entities related to the corresponding virtual entities.

- **IoT service organization functional group**: The purpose of the IoT service organization FG is to host all functional components that support the composition and orchestration of IoT and virtual entity services. Moreover, this FG acts as a service hub between several other functional groups, such as the IoT process management FG when, for example, service requests from applications or the IoT process management are directed to the resources implementing the necessary services.
- **IoT process management functional group**: The IoT process management FG is a collection of functionalities that allows smooth integration of IoT-related services (IoT services, virtual entity services, composed services) with the enterprise (business) processes.
- **Management functional group**: The management FG includes the necessary functions for enabling fault and performance monitoring of the system, configuration for enabling the system to be flexible to changing user demands, and accounting for enabling subsequent billing for the usage of the system. Support functions such as management of ownership, administrative domain, rules and rights of functional components, and information stores are also included in the management FG.
- **Security functional group**: The security FG contains the functional components that ensure the secure operation of the system as well as the management of privacy. The security FG contains components for the authentication of users (applications, humans), the authorization of access to services by users, secure communication (ensuring integrity and confidentiality of messages) between entities of the system such as devices, services, applications, and, last but not least, assurance of privacy of sensitive information relating to human users.
- **Application functional group**: The application FG is just a placeholder that represents all the needed logic for creating an IoT application. The applications typically contain custom logic tailored to a specific domain such as a smart grid.

2.6.1.4 Communication model

Safety: The IoT reference model can only provide IoT-related guidelines for ensuring a safe system to the extent possible and controllable by a system designer.

E.g.: smart grid.

Privacy: Because interactions with the physical world may often include humans, protecting user privacy is of the utmost importance for an IoT system. The IoT-A privacy model depends on the following functional components: identity management, authentication, authorization, and trust &reputation.

Trust: Generally, an entity is said to 'trust' a second entity when the first entity makes the assumption that the second entity will behave exactly as the first entity expects.

Security: The security model for IoT consists of communication security that focuses mostly on the confidentiality and integrity protection of interacting entities and functional components such as Identity Management, Authentication, Authorization, and Trust & Reputation.

There are several reasons why reference architecture for IoT is a good thing:

- IoT devices are inherently connected – we need a way of interacting with them, often with firewalls, network address translation (NAT) and other obstacles in the way.
- There are billions of these devices already and the number is growing quickly; we need architecture for scalability. In addition, these devices are typically interacting 24x7, so we need a highly-available (HA) approach that supports deployment across data centers to allow disaster recovery (DR).
- The devices may not have User Interfaces (UIs) and certainly are designed to be "everyday" usage, so we need to support automatic and managed updates, as well as being able to remotely manage these devices.
- IoT devices are very commonly used for collecting and analyzing personal data. A model for managing the identity and access control for IoT devices and the data they publish and consume is a key requirement.

Our aim is to provide an architecture that supports integration between systems and devices.

In the next section, we will dig into these requirements deeper and outline the specific requirements we are looking for in a range of categories.

2.6.2 IoT reference architecture

There are some specific requirements for IoT that are unique to IoT devices and the environments that support them, e.g. many requirements emerge from the limited-form factors and power available to IoT devices. Other requirements come from the way in which IoT devices are manufactured and used. The approaches are much more like traditional consumer product designs than existing Internet approaches. Of course, there are a number of existing best practices for the server-side and Internet connectivity that need to be remembered and factored in.

We can summarize the overall requirements into some key categories:

- Connectivity and communications
- Device management

- Data collection, analysis, and actuation
- Scalability
- Security
- HA
- Predictive analysis
- Integration

2.6.2.1 Connectivity and communications

Existing protocols, such as HTTP, have a very important place in many devices. Even an 8-bit controller can create simple GET and POST requests and HTTP provides an important unified (and uniform) connectivity. However, the overhead of HTTP and some other traditional Internet protocols can be an issue for two main reasons. Firstly, the memory size of the program can be an issue on small devices. However, the bigger issue is the power requirements. In order to meet these requirements, we need a simple, small and binary protocol. We will look at this in more detail below. We also require the ability to cross firewalls.

In addition, there are devices that connect directly and those that connect via gateways. The devices that connect via a gateway potentially require two protocols: one to connect to the gateway, and another from the gateway to the cloud.

Finally, there is obviously a requirement for our architecture to support transport and protocol bridging, e.g. we may wish to offer a binary protocol to the device, but allow an HTTP-based API to control the device that we expose to third parties.

2.6.2.2 Device management

While many IoT devices are not actively managed, this is not necessarily ideal. We have seen active management of PCs, mobile phones, and other devices become increasingly important, and the same trajectory is both likely and desirable for IoT devices. What are the requirements for IoT device management? The following list covers some widely desirable requirements:

- The ability to disconnect a rogue or stolen device
- The ability to update the software on a device
- Updating security credentials
- Remotely enabling or disabling certain hardware capabilities
- Locating a lost device
- Wiping secure data from a stolen device
- Remotely re-configuring Wi-Fi, GPRS, or network parameters

The list is not exhaustive, and conversely covers aspects that may not be required or possible for certain devices.

2.6.2.3 Data collection, analysis, and actuation

A few IoT devices have some form of UI, but in general IoT devices are focused on offering one or more sensors, one or more actuators, or a combination of both. The requirements of the system are that we can collect data from very large numbers of devices, store it, analyze it, and then act upon it.

The reference architecture is designed to manage very large numbers of devices. If these devices are creating constant streams of data, then this creates a significant amount of data. The requirement is for a highly scalable storage system, which can handle diverse data and high volumes.

The action may happen in near real time, so there is a strong requirement for real-time analytics. In addition, the device needs to be able to analyze and act on data. In some cases this will be simple, embedded logic. On more powerful devices we can also utilize more powerful engines for event processing and action.

2.6.2.4 Scalability

Any server-side architecture would ideally be highly scalable, and be able to support millions of devices all constantly sending, receiving, and acting on data. However, many "high-scalability architectures" have come with an equally high price – in hardware, software, and complexity. An important requirement for this architecture is to support scaling from a small deployment to a very large number of devices. Elastic scalability and the ability to deploy in a cloud infrastructure are essential. The ability to scale the server side out on small cheap servers is an important requirement to make this an affordable architecture for small deployments as well as large ones.

2.6.2.5 Security

Security is one of the most important aspects for IoT. IoT devices are often collecting highly personal data, and by their nature they are bringing the real world onto the Internet (and vice versa). This brings three categories of risks:

- Risks that are inherent in any Internet system, but that product/IoT designers may not be aware of
- Specific risks that are unique to IoT devices
- Safety to ensure no harm is caused by, for instance, misusing actuators

The first category includes simple things such as locking down open ports on devices (like the Internet-attached fridge that had an unsecured SMTP server and was being used to send spam).

The second category includes issues specifically related to IoT hardware, e.g. the device may have its secure information read. For example, many IoT devices are too small to support proper asymmetric encryption. Another specific example is the ability for someone to attack the hardware to understand security. Another example – university security researchers who famously reverse-engineered and broke the Mifare Classic RFID card solution. These sort of reverse engineering attacks are an issue compared with pure web solutions where there is often no available code to attack (i.e. completely server-side implementation).

Two very important specific issues for IoT security are (1) the concerns about identity and (2) access management. Identity is an issue where there are often poor practices implemented. For example, the use of clear text/ Base64 encoded user IDs/passwords with devices and machine-to-machine (M2M) is a common mistake. Ideally, these should be replaced with managed tokens such as those provided by OAuth/OAuth24.

Another common issue is to hard-code access management rules into either client- or server-side code. A much more flexible and powerful approach is to utilize models such as "Attribute Based Access Control" and "Policy Based Access Control".

Our security requirements therefore should support:

- Encryption on devices that are powerful enough
- A modern identity model based on tokens and not user IDs/passwords
- The management of keys and tokens as smoothly/remotely as possible and
- Policy-based and user-managed access control for the system based on XACML

This concludes the set of requirements that we have identified for the reference architecture. Of course, any given architecture may add further requirements. Some of those may already be met by the architecture, and some may require further components to be added. However, our design is for a modular architecture that supports extensions, which copes with this demand.

The reference architecture consists of a set of components. Layers can be realized by means of specific technologies, and we will discuss options for realizing each component. There are also some cross-cutting/vertical layers such as access/identity management (Figure 2.31).

The layers are:

- Client/external communications – Web/Portal, Dashboard, APIs
- Event processing and analytics (including data storage)
- Aggregation/bus layer – ESB and message broker
- Relevant transports – MQTT/HTTP/XMPP/CoAP/AMQP, etc.
- Devices

The cross-cutting layers are:

- Device manager
- Identity and access management

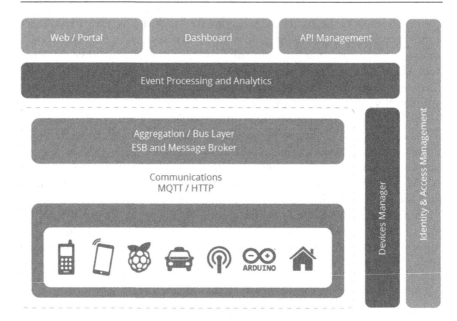

Figure 2.31 IoT reference architecture.

2.6.2.6 The device layer

The bottom layer of the architecture is the device layer. Devices can be of various types, but in order to be considered as IoT devices, they must have some communications that either indirectly or directly attaches to the Internet. Examples of direct connections are:

- Arduino with Arduino Ethernet connection
- Arduino Yun with a Wi-Fi connection
- Raspberry Pi connected via Ethernet or Wi-Fi
- Intel Galileo connected via Ethernet or Wi-Fi Examples of indirectly connected device include
- ZigBee devices connected via a ZigBee gateway
- Bluetooth or Bluetooth Low Energy devices connecting via a mobile phone
- Devices communicating via low power radios to a Raspberry Pi

There are many more such examples of each type. Each device typically needs an identity. The identity may be one of the following

- A unique identifier (UUID) burnt into the device (typically part of the System-on-Chip, or provided by a secondary chip)
- A UUID provided by the radio subsystem (e.g. Bluetooth identifier, Wi-Fi MAC address)

- An OAuth2 Refresh/Bearer Token (this may be in addition to one of the above)
- An identifier stored in nonvolatile memory such as EEPROM

For the reference architecture we recommend that every device has a UUID (preferably an unchangeable ID provided by the core hardware) as well as an OAuth2 Refresh and Bearer token stored in EEPROM.

The specification is based on HTTP; however, (as we will discuss in the communications section) the reference architecture also supports these flows over MQTT.

2.6.2.7 The communications layer

The communication layer supports the connectivity of the devices. There are multiple potential protocols for communication between the devices and the cloud. The most well-known three potential protocols are:

- HTTP/HTTPS (and RESTful approaches on those)
- MQTT 3.1/3.1.1
- Constrained application protocol (CoAP)

Let's take a quick look at each of these protocols in turn.

HTTP is well known, and there are many libraries that support it. Because it is a simple text-based protocol, many small devices, such as 8-bit controllers, can only partially support the protocol – for example, enough code to POST or GET a resource. The larger 32-bit based devices can utilize full HTTP client libraries that properly implement the whole protocol.

There are several protocols optimized for IoT use. The two best-known ones are MQTT6 and CoAP7. MQTT was invented in 1999 to solve issues in embedded systems and SCADA. It has been through some iterations and the current version (3.1.1) is undergoing standardization in the OASIS MQTT Technical Committee 8. MQTT is a publish–subscribe messaging system based on a broker model. The protocol has a very low overheads (as little as 2 bytes per message), and was designed to support lossy and intermittently connected networks. MQTT was designed to flow over TCP. In addition, there is an associated specification designed for ZigBee-style networks called MQTT-SN (Sensor Networks).

CoAP is a protocol from the IETF that is designed to provide a RESTful application protocol modeled on HTTP semantics, but with a much smaller footprint and a binary rather than a text-based approach. CoAP is a more traditional client–server approach rather than a brokered approach. CoAP is designed to be used over UDP.

For the reference architecture we have opted to select MQTT as the preferred device communication protocol, with HTTP as an alternative option.

The reasons to select MQTT and not CoAP at this stage are:

- Better adoption and wider library support for MQTT
- Simplified bridging into existing event collection and event-processing systems and
- Simpler connectivity over firewalls and NAT networks

However, both protocols have specific strengths (and weaknesses) and so there will be some situations where CoAP may be preferable and could be swapped in.

In order to support MQTT we need to have an MQTT broker in the architecture as well as device libraries. We will discuss this with regard to security and scalability later.

One important aspect with IoT devices is not just for the device to send data to the cloud/server, but also for the reverse to be possible. This is one of the benefits of the MQTT specification: because it is a brokered model, clients connect an outbound connection to the broker, whether or not the device is acting as a publisher or subscriber. This usually avoids firewall problems because this approach works even behind firewalls or via NAT.

In the case where the main communication is based on HTTP, the traditional approach for sending data to the device would be to use HTTP Polling. This is very inefficient and costly, in terms of both network traffic as well as power requirements. The modern replacement for this is the Web Socket protocol which allows an HTTP connection to be upgraded into a full two-way connection. This then acts as a socket channel (similar to a pure TCP channel) between the server and client. Once that has been established, it is up to the system to choose an ongoing protocol to tunnel over the connection.

For the reference architecture we once again recommend using MQTT as a protocol with Web Sockets. In some cases, MQTT over Web Sockets will be the only protocol. This is because it is even more firewall-friendly than the base MQTT specification as well as being able to support pure browser/ JavaScript clients using the same protocol.

Note that while there is some support for Web Sockets on small controllers, such as Arduino, the combination of network code, HTTP and Web Sockets would utilize most of the available code space on a typical Arduino 8-bit device. Therefore, we only recommend the use of Web Sockets on the larger 32-bit devices.

2.6.2.8 The aggregation/bus layer

An important layer of the architecture is the layer that aggregates and brokers communications. This is an important layer for three reasons:

1. The ability to support an HTTP server and/or an MQTT broker to talk to the devices

2. The ability to aggregate and combine communications from different devices and to route communications to a specific device (possibly via a gateway)
3. The ability to bridge and transform between different protocols, e.g. to offer HTTP based APIs that are mediated into an MQTT message going to the device

The aggregation/bus layer provides these capabilities as well as adapting into legacy protocols. The bus layer may also provide some simple correlation and mapping from different correlation models (e.g. mapping a device ID into an owner's ID or viceversa).

Finally, the aggregation/bus layer needs to perform two key security roles. It must be able to act as an OAuth2 Resource Server (validating Bearer Tokens and associated resource access scopes). It must also be able to act as a policy enforcement point (PEP) for policy-based access. In this model, the bus makes requests to the identity and access management layer to validate access requests. The identity and access management layer acts as a policy decision point (PDP) in this process. The bus layer then implements the results of these calls to the PDP to either allow or disallow resource access.

2.6.2.9 The event processing and analytics layer

This layer takes the events from the bus and provides the ability to process and act upon these events. A core capability here is the requirement to store the data into a database. This may happen in three forms. The traditional model here would be to write a server-side application, e.g. this could be a JAX-RS application backed by a database. However, there are many approaches where we can support more agile approaches. The first of these is to use a Big Data analytics platform. This is a cloud-scalable platform that supports technologies such as Apache Hadoop to provide highly scalable map reduce analytics on the data coming from the devices. The second approach is to support complex event processing to initiate near real-time activities and actions based on data from the devices and from the rest of the system.

Our recommended approach in this space is to use the following approaches:

- Highly scalable, column-based data storage for storing events
- Map-reduce for long-running batch-oriented processing of data
- Complex event processing for fast in-memory processing and near real-time reaction and autonomic actions based on the data and activity of devices and other systems
- In addition, this layer may support traditional application processing platforms, such as Java Beans, JAX-RS logic, message-driven beans, or alternatives, such as node.js, PHP, Ruby or Python

2.6.2.10 Client/external communications layer

The reference architecture needs to provide a way for these devices to communicate outside of the device-oriented system. This includes three main approaches. Firstly, we need the ability to create web-based frontends and portals that interact with devices and with the event-processing layer. Secondly, we need the ability to create dashboards that offer views into analytics and event processing. Finally, we need to be able to interact with systems outside this network using machine-to-machine communications (APIs). These APIs need to be managed and controlled and this happens in an API management system.

The recommended approach to building the web front end is to utilize a modular front-end architecture, such as a portal, which allows simple fast composition of useful UIs. Of course, the architecture also supports existing Web server-side technology, such as Java Servlets/JSP, PHP, Python, Ruby, etc. Our recommended approach is based on the Java framework and the most popular Java-based web server, Apache Tomcat.

The dashboard is a re-usable system focused on creating graphs and other visualizations of data coming from the devices and the event-processing layer.

The API management layer provides three main functions:

- The first is that it provides a developer-focused portal (as opposed to the user-focused portal previously mentioned), where developers can find, explore, and subscribe to APIs from the system. There is also support for publishers to create, version, and manage the available and published APIs;
- The second is a gateway that manages access to the APIs, performing access control checks (for external requests) as well as throttling usage based on policies. It also performs routing and load-balancing;
- The final aspect is that the gateway publishes data into the analytics layer where it is stored as well as processed to provide insights into how the APIs are used.

2.6.2.11 Device management

Device management (DM) is handled by two components. A server-side system (the device manager) communicates with devices via various protocols and provides both individual and bulk control of devices. It also remotely manages software and applications deployed on the device. It can lock and/ or wipe the device if necessary. The device manager works in conjunction with the device management agents. There are multiple different agents for different platforms and device types.

The device manager also needs to maintain the list of device identities and map these into owners. It must also work with the identity and access management layer to manage access controls over devices (e.g. who else can manage the device apart from the owner, how much control does the owner have vs. the administrator, etc.)

There are three levels of device: non-managed, semi-managed, and fully managed (NM, SM, FM).

Fully managed devices are those that run a full DM agent. A full DM agent supports:

- Managing the software on the device
- Enabling/disabling features of the device (e.g. camera, hardware, etc.)
- Managing security controls and identifiers
- Monitoring the availability of the device
- Maintaining a record of the device's location if available
- Locking or wiping the device remotely if the device is compromised, etc.

Non-managed devices can communicate with the rest of the network, but have no agent involved. These may include 8-bit devices where the constraints are too small to support the agent. The device manager may still maintain information on the availability and location of the device if this is available.

Semi-managed devices are those that implement some parts of the DM (e.g. feature control, but not software management).

2.6.2.12 Identity and access management

The final layer is the identity and access management layer. This layer needs to provide the following services:

- OAuth2 token issuing and validation
- Other identity services, including SAML2 SSO and OpenID Connect support for identifying inbound requests from the Web layer
- XACML PDP
- Directory of users (e.g. LDAP)
- Policy management for access control (policy control point)

The identity layer may, of course, have other requirements specific to the other identity and access management for a given instantiation of the reference architecture. In this section we have outlined the major components of the reference architecture as well as specific decisions we have taken around technologies. These decisions are motivated by the specific requirements of IoT architectures as well as best practices for building agile, evolvable, scalable Internet architectures. Of course, there are other options, but this reference architecture utilizes proven approaches that are known to be successful in real-life IoT projects we have worked on.

2.7 MAPPING TO THE WSO2 PLATFORM

Reference architecture is useful as it is. However, it is even more useful if there is a real instantiation. In this section we provide a mapping into products and capabilities of the WSO2 platform to show how the reference architecture can be implemented.

The WSO2 platform is a completely modular, open-source enterprise platform that provides all the capabilities needed for the server side of this architecture. In addition, we also provide some reference components for the device layer – it is an intractable problem to provide components for all possible devices, but we do provide either sample code and/or supported code for certain popular device types.

An important aspect of the WSO2 platform is that it is inherently multi-tenant. This means that it can support multiple organizations on a single deployment with isolation between organizations (tenants). This is a key capability for deploying this reference architecture as a Software-as-a-Service (SaaS) offering. It is also used by some organizations on-premise to support different divisions or departments within a group.

The WSO2 platform supports deployment on three different targets:

1. Traditional on-premise servers, including Linux, Windows, Solaris, and AIX
2. Public cloud deployment, including Amazon EC2, Microsoft Azure, and Google Compute Engine
3. Hybrid or private cloud deployment on platforms, including OpenStack, Suse Cloud, Eucalyptus, Amazon Virtual Private Cloud, and Apache Stratos

The WSO2 platform is based on a technology called WSO2 Carbon, which is in turn based on OSGi. Each product in the platform shares the same kernel based on Carbon. In addition, each product is made from features that are composed to provide the required functionality. Features can be added and subtracted as needed. All the products work together using standard interoperable protocols, such as HTTP, MQTT, and AMQP. All the WSO2 products are available under the Apache Software License v2.0, which is a business friendly, non-viral Open Source License. Figure 2.32 shows the IoT reference architecture layered with the corresponding WSO2 product capabilities.

2.7.1 The device layer

We support any device. We have a reference device management capability on any Linux-based or Android-based device, which can be ported to other

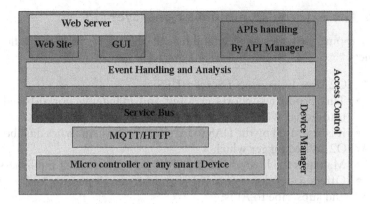

Figure 2.32 WSO2 platform.

32-bit platforms.WSO2 also can help with MQTT client code for many device platforms, ranging from Arduino to Android.

2.7.2 The aggregation/bus layer

We provide two core products that implement this layer:

- WSO2 Enterprise Service Bus (ESB), which provides HTTP, MQTT, AMQP and other protocol support, protocol mediation and bridging, data transformation, OAuth2 Resource Server support, XACML Policy Enforcement Point (PEP) support, and many other capabilities. WSO2 ESB is highly scalable, providing linear scalability and elastic scalability. In one deployment it handles more than 2bn requests/day. Please note that WSO2 ESB does not currently support Web Sockets, but this is on the roadmap.
- WSO2 Message Broker (MB), which provides the ability to act as an MQTT broker. WSO2 MB also provides AMQP capabilities and can provide both persistent and non-persistent messaging. WSO2 MB is highly scalable and supports elastic scalability. Please note that the WSO2 MB MQTT support is currently in beta.

2.7.3 The analytics and event processing layer

WSO2 offers a complete platform for data analytics with WSO2 Data Analytics Server, an industry first that combines the ability to analyze the same data at rest and in motion with predictive analysis. WSO2 DAS replaces WSO2 Business Activity Monitor and WSO2 Complex Event Processor. WSO2's analytics platform offers the flexibility to scale to millions of events per second, whether running on-premises and in the cloud.

2.7.4 The external communications layer

Our mapping provides the capabilities of this layer with the following products:

- WSO2 User Engagement Server (UES)
  - This product supports creating and managing portal-based and traditional Web UIs, including supporting full personalization.
  - It is also used by the DAS to manage and host analytics dashboards.
- WSO2 API Manager which
  - Manages the lifecycle of the APIs and supports API publishers;
  - Offers a developer-focused portal for developers to find, explore and subscribe to APIs;
  - Issues and manages OAuth2 tokens to external developers (note that when WSO2 Identity Server is also available – see below – then this function is delegated to that system);
  - Gateways external requests and provides throttling and PEP capabilities;
  - Publishes usage, version, and other data into the DAS; and
  - Integrates with WSO2 ESB

2.7.5 The device management layer

WSO2 Enterprise Mobility Management (EMM) provides:

- Mobile device management for iOS, Android, and IoT devices
- A full app store for managing applications and provisioning applications on to managed devices
- Integration with the identity layer as well as DAS for mobile analytics

2.7.6 The identity and access management layer

WSO2 Identity Server supports this aspect, and provides the following capabilities:

- OAuth2 identity provider, issuing, revoking and managing tokens
- Single sign-on support, including SAML2 SSO and Open ID Connect support
- Support for other identity protocols, including WS-Federation (Passive), Open ID 2.0, Kerberos, Integrated Windows Authentication (IWA), and others
- Full support for XACML (including versions 2.0 and 3.0), acting as a PDP, PIP, and PAP
- The ability to integrate between different identity providers and service providers, including identity brokering
- Support for identity provisioning, including SPML and SCIM support

The WSO2 platform is the only modular, open-source platform to provide all these capabilities (and more). As such, it is the ideal basis for creating and deploying this IoT reference architecture.

One further aspect that is highly worthy of consideration is the use of a Platform-as-a-Service (PaaS). WSO2 provides the WSO2 Private PaaS product which is based on the Apache Stratos project. This provides a managed, elastically scalable, HA deployment of the products mentioned above and also manages tenancy, self-service subscription, and many other aspects. It also supports managing many other useful server-side capabilities, including PHP, MySQL, MongoDB and others. We have not shown the PaaS layer on the IoT reference architecture as some deployments may not need this capability.

2.8 DESIGN PRINCIPLES FOR IoT

In this part, the focus is on design principles of IoT that one needs to follow to design IoT products/services. These design principles allow the developers to analyze and implement the IoT:

Interoperability: The IoT is going to develop in a number of areas and it may find that in a number of applications its very diversity can be the main obstacle to its own growth. The devices that will be forming the IoT will be countless and of different types i.e. varying technical profiles will operate (from household appliances to wearable, from autonomous vehicles to drones, and so on), manufactured by thousands of different companies, each with their own standards. It is basically the ability for systems or components of systems to communicate with each other, regardless of their manufacturer or technical specifications. Imagine that you are travelling in an autonomous vehicle and you need to communicate with other vehicles on the road to co-ordinate your movements and to be able to drive safely. What if they could not do so because each vehicle is manufactured by a different company which would make the exchange of information impossible? In this type of situation, even people's lives could be put at risk. This makes interoperability an important principle of IoT development.

Virtualization: Devices or systems must be able to simulate and create a virtual copy of real world. For example, imagine a factory environment, with nodes deployed for temperature monitoring & control, or machine monitoring & control and a central head controlling the entire factory. In the case of a slight or obvious temperature rise (of factory/machinery), the nodes should be automatically able to control temperature (let's say by changing thermostat's temperature/coolant flow) and then report to the central head later, since it was not a huge deal and didn't need immediate attention.

Decentralization: This is the ability of each device to work independently, in case of issues or the absence of a controlling master.

Real-Time Capability: An IOT device/system needs to be able to collect data, store or analyse it, and make decisions according to new findings in real time.

Service Orientation: Products must be customer-oriented. People and smart objects/devices must be able to connect efficiently through the Internet of Services to create products based on the customer's specifications. E.g. One might make an IOT application but find that it is not useful to any customer as it doesn't satisfy any of their needs or address any of their problems.

Modularity: an IOT device's ability to adapt to a new market is essential.

2.9 IoT AND M2M TECHNOLOGY

IoT M2M(machine-to-machine) verbal exchange involves two machines "speaking," or exchanging information, without human interaction. This consists of serial connection, power line connection (percent), or Wi-Fi communications inside the business Internet of Things (IoT).

Switching over to Wi-Fi has made the M2M verbal exchange a lot simpler and enabled greater packages to be relayed.

In widespread, when a person says M2M communication, they regularly are regarding cellular communication for embedded gadgets.

Examples of M2M communication in this example would be vending machines sending out inventory records or ATM machines obtaining the authorization to dispense cash.

2.9.1 How IoT M2M works

As previously stated, device-to-system communicate makes the Internet of Things possible. In step with Forbes, M2M is among the fastest-growing forms of connected tool technologies inside the marketplace proper now, largely due to the fact M2M technology can connect millions of gadgets inside an unmarried community. The variety of linked gadgets includes whatever from vending machines to medical device to automobiles to homes (Figure 2.33).

This sounds complex, but the driving concept behind the idea is pretty simple. Essentially, M2M networks are very just like LAN or WAN networks, but are solely used to allow machines, sensors, and controls, to communicate.

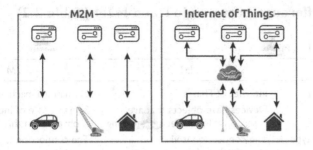

Figure 2.33 IoT and M2M.

Figure 2.34 IoT M2M applications.

Those gadgets feed facts they acquire returned to other gadgets inside the network. This system lets in a human to evaluate what is going on across the complete network and issue suitable instructions to member gadgets (Figure 2.34 and Table 2.2).

2.9.2 IoT M2M applications

M2M systems use point-to-point communications between machines, sensors and hardware over cellular or wired networks, while IoT systems rely on IP-based networks to send data collected from IoT-connected devices to gateways, the cloud or middleware platforms

2.9.3 Difference between IoT and M2M (Table 2.2)

Table 2.2 IoT vs M2M

Basis of	IoT	M2M
Abbreviation	Internet of Things	Machine to machine
Intelligence	Devices have objects that are responsible for decision-making	Some degree of intelligence is observed in this
Connection type used	The connection is via network and using various communication types.	The connection is point to point
Communication protocol used	Internet protocols are used such as HTTP, FTP, and Telnet.	Traditional protocols and communication technology techniques are used
Data Sharing	Data is shared between other applications that are used to improve the end-user experience.	Data is shared with only the communicating parties.
Internet	Internet connection is required for communication	Devices are not dependent on the Internet.
Scope	A large number of devices yet scope is large.	Limited scope for devices.
Business type used	Business 2 Business (B2B) and Business 2 Consumer (B2C)	Business 2 Business (B2B)
Open API support	Supports Open API integrations.	There is no support for Open APIs
Examples	Smart wearable's, Big Data and Cloud, etc.	Sensors, data and information, etc.

2.10 REAL-WORLD DESIGN CONSTRAINTS

2.10.1 Devices and networks

The devices that form networks in the M2M area network domain must be selected, or designed, with certain functionality suitable to IoT applications. The devices must have an energy source (e.g. batteries), computational capability (e.g. an MCU), appropriate communications interface (e.g. a Radio Frequency Integrated Circuit (RFIC) and front-end RF circuitry), memory (program and data), and sensing (and/or actuation) capability. These must be integrated in such a way that the functional requirements of the desired application can be satisfied with additional nonfunctional requirements.

2.10.1.1 Functional requirements

1. Specific sensing and actuating capabilities
2. Sensing principle and data requirements: Sometimes continuous sampling of sensing data is required. For some applications, sampling after specific intervals is required.
3. The parameters such as higher network throughput, data loss, energy use, etc. are decided based on sensing principle.

2.10.1.2 Sensing and communications field

The sensing field is to be considered for sensing in local area or distributed sensing. The distance between sensing points is also an important factor to be considered. The physical environment has an implication for the communications technologies selected and the reliability of the system in operation thereafter. Devices must be placed in close enough proximity to communicate. Where the distance is too great, routing devices may be necessary.

2.10.1.3 Programming and embedded intelligence

Devices in the IoT are heterogeneous, such as various computational architectures including MCUs (8-, 16-, 32-bit, ARM, 8051, RISC, Intel, etc.), signal conditioning (e.g. ADC), and memory (ROM, S/F/D) RAM, etc.), communications media, peripheral components (sensors, actuators, buttons, screens, LEDs), etc. In every case, an application programmer must consider the hardware selected or designed, and its capabilities.

Application-level logic decides the sampling rate of the sensor, the local processing performed on sensor readings, the transmission schedule (or reporting rate), and the management of the communications protocol stack, among other things. The programmers have to reconfigure and reprogram devices in case of changes in the devices in IoT applications.

2.10.1.4 Power

Power is essential for any embedded or IoT device. Depending on the application, power may be provided by the mains, batteries, or hybrid power sources. Power requirements of the application are modeled prior to deployment. This allows the designer to estimate the cost of maintenance over time.

2.10.1.5 Gateway

Gateway devices or proxies are selected according to the needs of data transitions.

2.10.1.6 Nonfunctional requirements

The non-functional requirements are both technical and non-technical.

1. **Regulations:** For applications that require placing nodes in public places, prior permissions are important. Radio frequency (RF) regulations limit the power with which transmitters can broadcast.
2. **Ease of use, installation, maintenance, accessibility:** This relates to positioning, placement, site surveying, programming, and the physical accessibility of devices for maintenance purposes.
3. **Physical constraints:** Integration of additional electronics into existing system. These might include suitable packaging, the kind and size of antenna, the type of power supply etc.

2.10.1.7 Financial cost

Financial cost considerations are as follows:

- Component selection: Typically, the use of these devices in the M2M area network domain is to reduce the overall cost burden. However, there are research and development costs that are likely to be incurred for each individual application in the IoT which requires device development or integration. Developing devices in small quantities is expensive.
- Integrated device design: Once the energy, sensors, actuators, computation, memory, power, connectivity, physical, and other functional and non-functional requirements are considered, it is likely that an integrated device must be produced.

2.10.1.8 Data representation and visualization

Each IoT application has an optimal visual representation of the data and the system. Data that are generated from heterogeneous systems have heterogeneous visualization requirements. There are currently no satisfactory standard data representation and storage methods that satisfy all of the potential IoT applications.

Chapter 3

Protocols in IoT

As discussed in the previous chapter, protocols are used to establish communication between anode device and a server over the Internet. They help to send commands to an IoT device and receive data from an IoT device over the Internet. People use different types of protocols that are present on both the server and the client side and these protocols are managed by network layers like application, transport, network, and link layers.

3.1 MQTT (MESSAGE QUEUING TELEMETRY TRANSPORT)

MQTT is a machine-to-machine (M2M), Internet of Things (IoT) connectivity protocol. It was designed as an extremely lightweight publish/subscribe messaging transport protocol and is useful for connections with remote locations where a small code footprint is required and/or network bandwidth is at a premium. It makes it easy for communication between multiple devices. It is a simple messaging protocol designed for the constrained devices and with low bandwidth, so it's a perfect solution for the internet of things applications.

3.1.1 Why was MQTT created?

MQTT was created with the goal of collecting data from many devices and then transporting that data to the IT infrastructure. It is lightweight, and therefore ideal for remote monitoring, especially in M2M connections which require a small code footprint or where the network bandwidth is limited. MQTT was invented in 1999 by Dr. Andy Stanford-Clark and Arlen Nipper. The co-inventor Nipper is the president of Cirrus Link Solutions, the company which developed the Cirrus Link MQTT Modules for Ignition.

DOI: 10.1201/9781003307488-3

3.1.2 Who uses MQTT?

MQTT was originally developed for the low-bandwidth, high-latency data links used in the oil and gas industry. However, MQTT is now used in many applications beyond this sector, from controlling smart lighting systems to Facebook Messenger and Amazon Web Services (AWS) IoT. Overall, MQTT appears to be the protocol best suited for the control systems used by industrial organizations and we can expect that its rapid rate of adoption will only increase in future.

3.1.3 Advantages of MQTT

The MQTT protocol allows your supervisory control and data acquisition (SCADA) system to access Industrial Internet of Things (IIoT) data. MQTT brings many powerful benefits to your process:

- It distributes information more efficiently
- It increases scalability
- It reduces network bandwidth consumption dramatically
- It reduces update rates to seconds
- It is very well-suited for remote sensing and control
- It maximizes available bandwidth
- It has extremely low overheads
- It is very secure, with permission-based security
- It is used by the oil and gas industry, Amazon, Facebook and other major businesses
- It saves development time
- Publish/subscribe protocols collects more data with less bandwidth than polling protocols

3.2 HOW IT WORKS

To understand the MQTT architecture, we first look at the components of the MQTT.

- Message
- Client
- Server or broker
- TOPIC

Message: The message is the data that is carried out by the protocol across the network for the application. When the message is transmitted over the network, then the message contains the following parameters:

1. Payload data
2. Quality of Service (QoS)
3. Collection of Properties
4. Topic Name

Client: In MQTT, the client has two roles: as subscriber and publisher. The clients subscribe to the topics to publish and receive messages. In simple terms, we can say that if any program or device uses an MQTT, then that device is referred to as a client. A device is a client if it opens the network connection to the server, publishes messages that other clients want to see, subscribes to the messages that it is interested in receiving, unsubscribes from the messages that it is not interested in receiving, and closes the network connection to the server.

In MQTT, the client performs two operations:

Publish: When the client sends the data to the server, then we call this operation publishing.

Subscribe: When the client receives the data from the server, then we call this operation subscribing.

Server: This is the device or a program that allows the client to publish the messages and subscribe to the messages. A server accepts the network connection from the client, accepts the messages from the client, processes the subscribe and unsubscribe requests, forwards the application messages to the client, and closes the network connection from the client.

TOPIC: The label provided to the message is checked against the subscription known by the server is known as TOPIC.

To understand the process more clearly, we will consider a real-world example. Suppose a device has a temperature sensor and wants to send the rating to the server or the broker. If the phone or desktop application wishes to receive this temperature value on the other side, then two things will have happened. The publisher first defines the topic; for example, the temperature then publishes the message, i.e., the temperature's value as shown in Figure 3.1. After publishing the message, the phone or the desktop application on the other side will subscribe to the

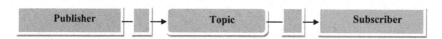

Figure 3.1 Publish subscribe model.

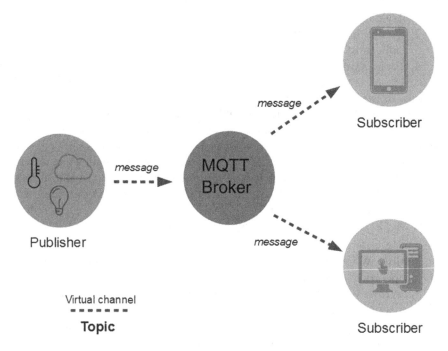

Figure 3.2 Function of broker.

topic, i.e., temperature, and then receive the published message, i.e., the value of the temperature. The server or the broker's role is to deliver the published message to the phone or the desktop application as shown in Figure 3.2.

3.2.1 Levels of QoS in MQTT

Quality of Service (QoS) defines the reliability of the message delivery process in MQTT. MQTT provides three QoS levels for message delivery: QoS 0, QoS 1, and QoS 2 as shown in Figure 3.3. You can have different QoS levels for publishing and for subscribing to messages. The MQTT broker that you are using might not support all three levels of QoS. For example, ThingSpeak MQTT supports only QoS 0.

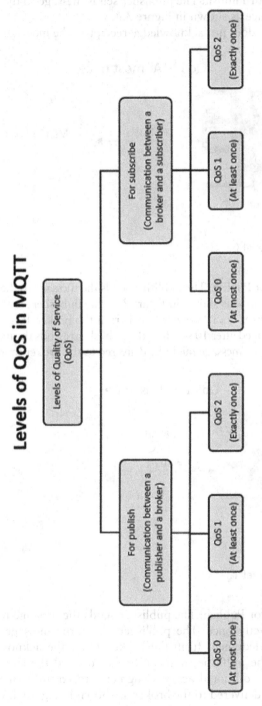

Figure 3.3 Levels of QoS.

QoS 0 Level for Publish: The publisher sends message to the MQTT broker only once as shown in Figure 3.4.
The broker does not acknowledge receipt of the message.

QoS 0: At most once

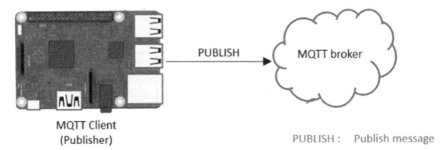

Figure 3.4 QoS Level 0.

QoS 1 Level for Publish: The publisher sends the message to the MQTT broker at least once as shown in Figure 3.5. The publisher stores the message until it receives an acknowledgment from the broker. If no acknowledgment is received after 10 seconds, the publisher resends the message. In this level, the same message might be delivered to the broker more than once.

QoS 1: At least once

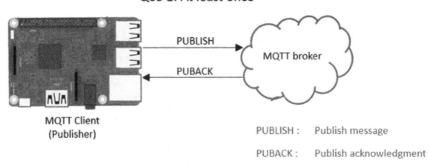

Figure 3.5 QoS Level I.

QoS 2 Level for Publish: The publisher sends the message to the MQTT broker exactly once. The publisher stores the message until it gets an acknowledgment from the broker. Once the acknowledgment is received, the publisher and the broker discard the stored messages. QoS2 uses additional acknowledgments to ensure that no duplicate message is delivered to the broker as shown in Figure 3.6.

QoS 2: Exactly once

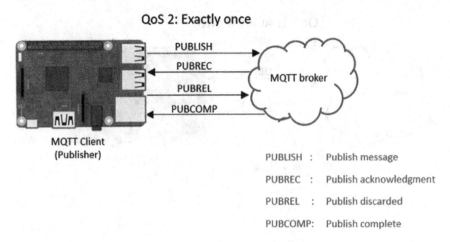

PUBLISH

PUBREC

MQTT broker

PUBREL

PUBCOMP

MQTT Client
(Publisher)

PUBLISH	:	Publish message
PUBREC	:	Publish acknowledgment
PUBREL	:	Publish discarded
PUBCOMP:		Publish complete

Figure 3.6 QoS Level 2.

QoS 0 Level for Subscribe: The MQTT broker sends the message to the client only once as shown in Figure 3.7. The client does not acknowledge receipt of the message.

QoS 0: At most once

MQTT broker   PUBLISH

MQTT Client
(Publisher)

PUBLISH :   Publish message

Figure 3.7 QoS Level 0 (for subscribe).

QoS 1 Level for Subscribe: The MQTT broker sends the message to the client at least once as shown in Figure 3.8. The MQTT broker stores the message until it gets an acknowledgment from the client. If no acknowledgment is received after 10 seconds, the broker resends the message. In this level, the same message might be delivered more than once.

QoS 1: At least once

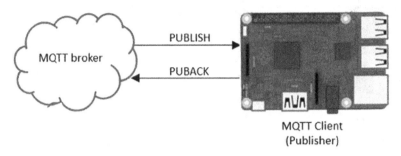

MQTT Client
(Publisher)

PUBLISH : Publish message

PUBACK : Publish acknowledgment

Figure 3.8 QoS Level I (for subscribe).

QoS 2 Level for Subscribe: The MQTT broker sends the message to the client exactly once as shown in Figure 3.9. The MQTT broker stores the message until it gets an acknowledgment from the client. Once the acknowledgment is received, the broker and the subscriber discard the stored messages. QoS2 uses additional acknowledgments to ensure that no duplicate message is delivered.

QoS 2: Exactly once

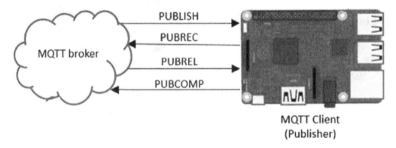

MQTT Client
(Publisher)

PUBLISH : Publish message

PUBREC : Publish acknowledgment

PUBREL : Publish discarded

PUBCOMP: Publish complete

Figure 3.9 QoS Level 2 (for subscribe).

3.3 ZIGBEE

Zigbee is a short-range, low-power, low-data rate wireless protocol used primarily for home automation and industrial control. In contrast to Wi-Fi networks used to connect endpoints to high-speed networks, Zigbee supports much lower data rates and uses a mesh networking protocol to avoid hub devices and create a self-healing architecture.

3.3.1 Zigbee Alliance

The Zigbee Alliance works to simplify wireless product integration in order to help product manufacturers introduce energy-efficient wireless control into their products faster and more cost-effectively. Alliance members create standards that offer reliable, secure, low-power, and easy-to-use wireless communication, using an open standards development process to guide their work. The alliance is organized by committees, work groups, study groups, taskforces, and special interest groups.

There are three types of membership with different rights and benefits:

- An **adopter** offers access to final, approved specifications, participation in interoperabilityeventsandaccesstostandardwork/taskgroupdocuments anddevelopmentactivities.
- A **participant** offers voting rights in work groups and has early access to all Zigbee Alliance standards and specifications in development.
- A **promoter** offers automatic voting rights in all work groups, final approval rights on all standards and a seat on the alliance's board of directors.

3.3.2 Who uses Zigbee?

Zigbee is used by a variety of cable and telecommunication companies in their set-top boxes, satellite transceivers, and home gateways to provide home monitoring and energy management products to their customers. Zigbee is also used by vendors that provide connected lighting products for homes and businesses. With Zigbee-based smart home products, consumers can control LED figures, light bulbs, remote controls, and switches both at home and remotely to improve energy management. Utility companies can use Zigbee in their smart meters to monitor, control, inform, and automate the delivery and use of energy and water. Smart meters give the consumers the information and automation needed to reduce energy use and save money. Zigbee-based products also enhance the shopping experience for consumers by enabling faster checkouts, in-store assistance, and in-store item location. Zigbee helps retailers operate more efficiently by ensuring items don't run out of stock and supporting just-in-time inventory practices as well as monitoring temperatures, humidity, spills, and soon. Zigbee supports a number of devices, including intelligent shopping carts, personal shopping assistants, electronic shelf labels, and asset-tracking tags.

Application areas: Zigbee protocols are intended for embedded applications requiring low power consumption and tolerating low data rates. The resulting network will use very little power. Individual devices must have a battery life of at least two years to pass certification.

Typical application areas include:

- Home automation
- Wireless sensor networks
- Industrial control systems
- Embedded sensing
- Medical data collection
- Smoke and intruder warning
- Building automation
- Remote wireless microphone configuration

3.3.3 How Zigbee works

There are three classes of Zigbee devices:

- **Zigbee coordinator (ZC):** The most capable device, the coordinator forms the root of the network tree and may bridge to other networks. There is precisely one Zigbee coordinator in each network since it is the device that started the network originally (the Zigbee Light Link specification also allows operation without a Zigbee coordinator, making it more usable for off-the-shelf home products). It stores information about the network, including acting as the trust center and repository for security keys. It is responsible for forming a centralized network.
- **Zigbee router (ZR):** As well as running an application function, a router can act as an intermediate node between the coordinator and end devices, passing data on from other nodes. Zigbee router devices provide routing services to network nodes. Routers can also serve as end nodes. In contrast to end nodes, routers are not designed to sleep and should generally remain on as long as a network is established.
- **Zigbee end device (ZED):** It contains enough functionality to talk to the parent node (either the coordinator or a router); it cannot relay data from other devices. This relationship allows the node to be asleep a significant amount of the time thereby giving long battery life. A ZED requires the least amount of memory and thus can be less expensive to manufacture than a ZR or ZC.

3.3.4 Zigbee network topologies

The Zigbee network layer is responsible for the formation of the network. There are three Zigbee network topologies: star, cluster tree, and mesh. These three Zigbee topologies come under one of the two network topologies mentioned in the IEEE 802.15.4, i.e. star and peer-to-peer.

Star Topology: In a Star network configuration, there is one coordinator and any number of end devices as shown in Figure 3.10. All the end devices are connected to the coordinator and the individual end devices are isolated, both physically and electrically, i.e. no direct communication between end devices.

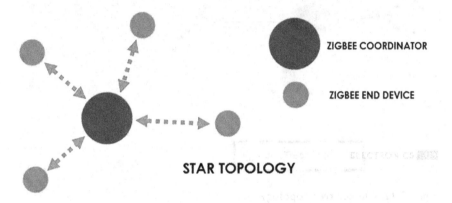

Figure 3.10 Star topology.

All the information must pass through the coordinator, even including the information from one end device to another as shown in Figure 3.11.

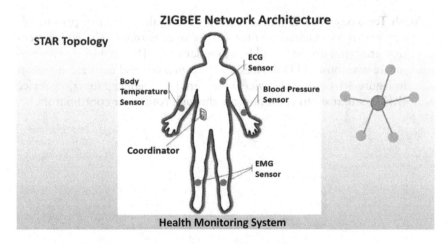

Figure 3.11 Zigbee architecture.

Cluster Tree Topology: This is a type of peer-to-peer topology. In this, the end devices join the network via the coordinator or the router as shown in Figure 3.12. As the Zigbee router extends the range of the Zigbee network, the end device need not be in the range of the coordinator.

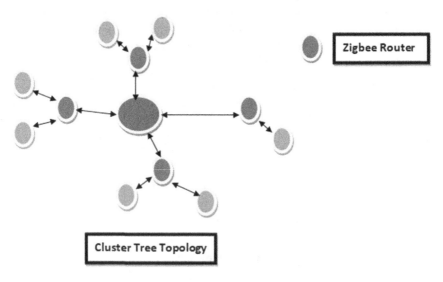

Figure 3.12 Cluster tree topology.

Even in tree topology, the end devices cannot communicate with each other directly, but the router can communicate with other routers and the coordinator.

Mesh Topology: The Zigbee mesh topology is also a peer-to-peer topology and is an extension to the cluster tree topology. The end devices that are configured as a full function device (FFD) can directly communicate with other FFD devices (either routers or end devices) as shown in Figure 3.13. But end devices configured as a reduced function device (RFD), still need to communicate through routers or coordinators.

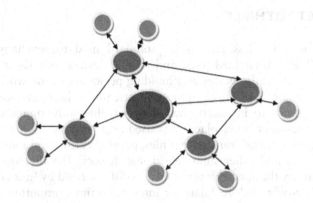

Figure 3.13 Mesh topology.

3.3.5 Applications of Zigbee technology

Zigbee networking and Zigbee technology have a wide range of applications, such as home automation, healthcare, and material tracking. Let us outline a few applications of Zigbee technology, where Zigbee devices can increase efficiency and reduce cost.

Home Automation

- Security Systems
- Meter Reading Systems
- Light Control Systems
- HVAC Systems

Consumer Electronics

- Gaming Consoles
- Wireless Mouse
- Wireless Remote Controls

Industrial Automation

- Asset Management
- Personnel Tracking
- Livestock Tracking
- Healthcare
- Hotel Room Access
- Fire Extinguishers

3.4 BLUETOOTH/BLE

Bluetooth is a wireless technology standard used for exchanging data between Bluetooth-enabled fixed and mobile devices over short distances using low-energy radio waves and building personal area networks (PANs) as shown in Figure 3.14. It is a connection-oriented technology, meaning that connection needs to be established between Bluetooth-compliant devices before data transfer takes place. It is used as a wireless alternative to RS-232 data cables. It can transfer text files, photos, videos, and more between mobile phones and other smart electronic devices. The Bluetooth project was initiated by the Special Interest Group (SIG) formed by four companies, IBM, Intel, Nokia and Toshiba, for interconnecting computing and communicating devices using short-range, low-powered, inexpensive wireless radios. As a result, it is also important for the rapidly growing Internet of Things, including smart homes and industrial applications. When Bluetooth devices connect to each other (for example, your phone and your wireless speaker), this follows the parent–child model, meaning that one device is the parent and other devices are the children. The parent transmits information to the child and the child listens for information from the parent.

A Bluetooth network is called a **piconet** and a collection of interconnected piconets is called a **scatternet**.

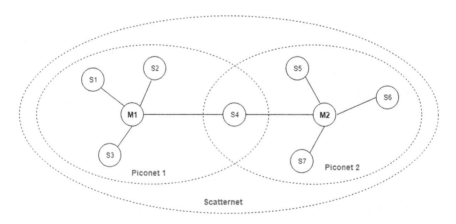

Figure 3.14 Bluetooth.

Piconet: This is a type of Bluetooth network that contains **one primary node** (called the master node) and **seven active secondary nodes** (called slave nodes). Thus, we can say that there are a total of 8 active nodes which are present at a distance of 10 meters. The communication between the primary and secondary nodes can be either one-to-one or one-to-many. Possible communication is only between the master and slave; Slave–slave communication is not possible. It also has

255 **parked nodes;** these are secondary nodes and cannot participate in communication unless it becomes converted to the active state.

Scatternet: This is formed by using **various piconets.** A slave that is present in one piconet can act as master (or we can say primary) in another piconet. This kind of node can receive a message from a master in one piconet and deliver the message to its slave in the other piconet where it is acting as a slave. This type of node is referred to as a bridge node. A station cannot be mastered in two piconets.

3.4.1 Applications of Bluetooth in IoT

Bluetooth Topologies: Pair, Broadcast, Mesh as shown in Figure 3.15:

Pair: Bluetooth as a means of pairing two devices. This might include, for example, a computer paired with a wireless mouse.

Broadcasting: Bluetooth as a means of having one device broadcast information to many devices or vice versa. This includes, for example, playing music on smart speakers and simultaneously casting photos to a projector, both using a single iPhone.

Mesh: Bluetooth as a way of connecting many devices to many others as if in a spider's web. For example, connecting 1,278 overhead lights in a warehouse to each other to dim and brighten lights automatically based upon activity and personal preferences.

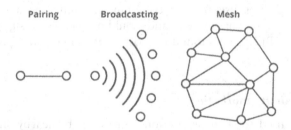

Figure 3.15 Bluetooth topologies.

Bluetooth is a useful technology for data transfers. It's now found in various devices and is popular for everything from music streaming to file sharing. One disadvantage of Bluetooth is that it can use a substantial amount of battery power on your device. If you leave a Bluetooth connection on all day, the additional power consumption is noticeable. This is particularly problematic for devices with limited power, such as those belonging to the Internet of Things, or even just your smartphone. Bluetooth Low Energy (BLE) has been designed to address this problem. So, what exactly is BLE and how does it work?

3.4.2 What is Bluetooth low energy?

Bluetooth Low Energy (BLE) is based on Bluetooth. It was released in 2011, and it is also referred to as either Bluetooth Smart or Bluetooth 4.0. BLE is designed to offer many of the same features as Bluetooth, but focusing on low power. As a result, it is not as fast as Bluetooth and is not suitable for transferring large files. It is, however, ideal for transferring small amounts of data with minimal power consumption. BLE has made it possible for a wide range of small IoT devices, such as sensors and tags, to communicate despite not having large batteries.

3.4.3 How does BLE use less power?

BLE uses the same radio wavebands as Bluetooth and allows two devices to exchange data in many of the same ways. The difference is that BLE devices remain asleep in between connections. They are also designed to only communicate for a few seconds when they do connect.

3.4.4 What is Bluetooth low energy used for?

BLE is never going to replace Bluetooth. But it has become the standard technology for many applications.

3.4.4.1 Smart devices

Most smart devices use BLE to communicate with one another. Many smart devices have limited power and would not be able to support normal Bluetooth use. BLE is also found in most smartphones, so it provides easy compatibility.

3.4.4.2 Proximity marketing

BLE can be used to send promotional messages to nearby smartphones. This allows marketing to be targeted to people based solely on their location. For example, a store might send notifications to people as they enter the premises.

3.4.4.3 Indoor location tracking

The Global Positioning System (GPS) is obviously effective in location tracking. However, it is not usually accurate enough to be used within very small areas such as inside buildings. BLE provides a useful alternative for indoor tracking. When combined with beacons, it can be used to track a smartphone from room to room.

3.4.4.4 Asset management

BLE can also be used to track physical items and is therefore popular in asset management. Each item to be tracked is given a BLE tag. Beacons are then set up throughout the premises to listen for the unique ID of each tag.

3.4.4.5 Is BLE the same as Bluetooth?

One more important question that we need to answer when talking about Bluetooth Low Energy is how it differs from Bluetooth Classic is shown in Figure 3.16 the technology we all know for its famous icon that we click on when we want to pair our devices. Does it differ at all?

The answer is yes. BLE is an independent standard which is incompatible with the classic Bluetooth. The latter was first introduced commercially over 20 years ago and is now essentially no longer being developed by the Bluetooth Special Interests Group (SIG). However, not being developed is not the same as "not being used." You will frequently find it in devices that require continuous connection, predominantly audio devices, such as wireless speakers or headphones.

Meanwhile, SIG introduced Bluetooth Low Energy in its 2010 Bluetooth 4.0 specification (with, later, the 2016 Bluetooth 5 specification, which was devoted exclusively to BLE). Its main focus was on the growing market of health- and fitness-related devices and smart home and indoor location.

	Bluetooth Classic	Bluetooth Low Energy
communication	continuous, bidirectional	short data transfers in one direction
range	100 m	<100 m
energy consumption	1 W	0.01–0.50 W
data rate	1–3 Mbit/s	125 kbit/s – 2 Mbit/s
latency	100 ms	6 ms
voice capable	yes	no

Figure 3.16 Bluetooth classic vs Bluetooth low energy.

3.4.4.6 Where can you find BLE?

Bluetooth Low Energy is used virtually anywhere, which is one of its core strengths compared to other low-powered networks. It is a commonly recognized and applied standard that essentially doesn't require specialized compatible hardware to be deployed. So if BLE is so popular, how is it being used?

3.4.4.7 Fitness trackers and smart appliances

This is one of the primary use cases for Bluetooth Low Energy, which arguably made it so ubiquitous. Because Bluetooth technologies (both classic and BLE) are so commonly available in smartphones, tablets, and laptops, it stands to reason that personal devices that we frequently pair with them – such as fitness trackers and various smart appliances – would use these, too. While you can think of even more constrained devices, smart appliances and trackers also usually have pretty heavy limitations. The bulk of their energy goes into basic functioning. Consider your smart band: it probably measures your steps or monitors your heartbeat continuously, using up a lot of battery life. If it were constantly relaying that information to your phone, too, it would drain the battery life a lot faster. This is why manufacturers look for ways to save energy here and there, and BLE is a perfect solution.

3.4.5 Indoor location tracking

One of the major BLE benefits is that it can be used for accurate positioning where GPS cannot be used indoors as shown in Figure 3.17. You can use BLE-equipped devices as beacons, i.e., to broadcast data to all devices in the vicinity, rather than to have one-to-one exchange. Based on that, devices capable of processing that data (such as phones) or simply capturing and relaying it further (such as access points) can determine the beacon's position.

Figure 3.17 Indoor location tracking.

This is why BLE is frequently used in indoor navigation systems, for example, in shopping malls that want to provide customers with GPS-like indoor mapping that will help them navigate to their favorite shop. But retail software for indoor positioning has a wider range of applications. Among other things, stores can use platforms such as Linkyfi to identify potential customers nearby and advertise their best deals (for example, via push notifications) for more targeted marketing that attracts more attention.

3.4.5.1 Contact tracing

Indoor positioning has gained even more importance when occupancy management became one of the top priorities for all businesses. To keep their colleagues safe, employers are increasingly looking into BLE-based solutions for unintrusive contact tracing as shown in Figure 3.18. In this scenario, people coming into the office are equipped with a simple, single-functioning BLE tag that collects the information on where they go and whom they come in contact with. At the same time, it doesn't store any sensitive data that might be considered an invasion of privacy. If someone in the office is sick, it is easy to identify who might have been exposed and contain the outbreak.

Solutions for location-based services, such as Linkyfi Location Engine, also allow offices to identify how people move around and where they tend to gather, providing heat maps of the most frequented places. This information can be used to manage occupancy and keep everyone safe; it can also be used to optimize space in general, such as by repurposing underused spaces.

Figure 3.18 Contact tracing.

3.4.5.2 Employee safety

BLE tags can help keep employees safe in more ways than one. Among the most common uses of this technology, other than contact tracing, are panic buttons and fall detection systems.

Panic buttons can be used by bank clerks, hotel staff, or anyone who works in a job that poses a degree of sudden danger. When carrying a BLE tag, these workers can discreetly call for help if they feel threatened, for example, by a suspicious customer. This will immediately alert the security and, thanks to location tracking, let them know the location of the person in danger.

With regard to the detection of a fall, BLE tags with movement sensors can be used in a range of locations, such as construction sites or nursing homes. Whenever the sensor registers a sudden fall, the BLE tag will automatically send an alert with the information of who's in danger and where they are so that they can be helped as quickly as possible.

3.4.5.3 Asset tracking

It's not only people that you can track with BLE tags. They can also be used for asset tracking as shown in Figure 3.19 simply attach the tag to objects you want to keep an eye on. You can either monitor how they move or, with geo-fencing, make sure they stay where they should be (e.g. carts in a shopping mall). This has applications for various verticals. For example, in logistics, it can be used to follow cargo. In medicine to monitor essential supplies so that they can always be easily found when needed. In IT – to track equipment to determine how, where, and when it's being used and then introduce optimizations, or to make sure it stays in the office. And these are just a few examples. It's applications like these, where the longevity and small data transfer that BLE allows are becoming crucial differentiators.

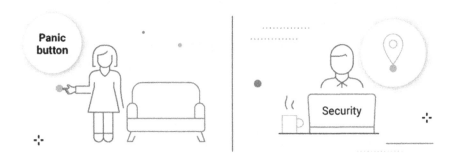

Figure 3.19 Asset tracking.

3.4.6 Architecture of BLE

Bluetooth Low Energy architecture is also described as the Bluetooth LE protocol stack as shown in Figure 3.20. This describes the different parts of the Bluetooth LE system, their components, and how they interact to yield the expected results. The BLE protocol stack architecture comprises three parts: the application layer, the host layer, and the controller layer.

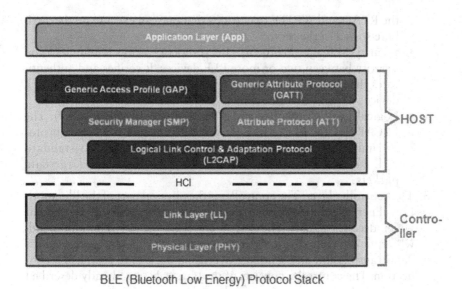

BLE (Bluetooth Low Energy) Protocol Stack

Figure 3.20 BLE architecture.

1. **The application layer**: This is the part that interacts directly with the user. It contains the user interface, application logic, and general application architecture. Underneath this layer is the actual hardware, which comprises the host and the controller layers.
2. **The host layer**: The host layer follows the application layer. It consists of various structures:
   - **Generic Access Profile (GAP)**: The GAP is a part of BLE architecture that describes how BLE devices communicate with each other. It includes peripheral or broadcaster devices, advertising information packets, and central device scanning for connection-ready devices.
   - **Generic Attribute Profile (GATT)**: This operates in a similar way to the GAP. It describes how attributes are formatted, packaged, and transferred across connected devices following a set of rules. The devices communicate as a client or a server. The client sends requests to the GATT server, which stores the attributes and makes them available on request. The client can either READ or WRITE or perform both functions on the attribute (data).
   - **Attribute Protocol**: The attribute protocol lays the foundation for the GATT profile to function. It is a set of rules guiding how data is accessed. It defines the GATT protocol's client–server rules, stating that a device can be a client, server, or function as both. The attribute protocol also defines the arrangement of data in the form of attributes, each having a 16-bit attribute handle, a universal unique identifier (UUID), a value, and a set of permissions. It also defined

the READ and WRITE operations that one can execute on the attribute stored in the server.

- **Security Manager Protocol**: This protocol ensures communication security between two or more BLE devices. It verifies and authenticates the pairing process. It can also prevent harmful tracking of a device's Bluetooth address by hiding it.
- **Logical Link Controller and Adaptation Protocol (L2CAP)**: The L2CAP is vital to the BLE architecture. It functions as a protocol multiplexer by converting multiple protocols into standard BLE packets. It can also break down and recombine large data packets.

3. **The controller layer**: The controller is the physical part of the Bluetooth Low Energy architecture hardware component. It holds the circuit which decodes signals. The chip operates on the 2.4GHz radio band, which it effectively divides into 40 channels. The channels are used for data transmission and sending advertising packets to establish a connection. The controller consists of the physical layer already described and the link layer which scans, advertises, creates, and monitors communication between BLE devices.

3.5 HTTP

The Hypertext Transfer Protocol (HTTP) is an application protocol for distributed, collaborative, hypermedia information systems that allows users to communicate data on the World Wide Web as shown in Figure 3.21. HTTP was invented alongside HTML to create the first interactive, text-based web browser: the original World Wide Web. Today, the protocol remains one of the primary means of using the Internet. Hypertext is structured text that uses logical links (hyperlinks) between nodes containing text. HTTP is the protocol to exchange or transfer hypertext. The standards development of HTTP was coordinated by the Internet Engineering Task Force (IETF) and the World Wide Web Consortium (W3C), culminating in the publication of a series of Requests for Comments (RFCs).

HTTP data rides above the TCP protocol, which guarantees the reliability of delivery, and breaks down large data requests and responses into network-manageable chunks. TCP is a "connection"-oriented protocol, which means when a client starts a dialogue with a server the TCP protocol will open a connection, over which the HTTP data will be reliably transferred. When the dialogue is complete that connection should be closed. All of the data in the HTTP protocol is expressed in human-readable ASCII text.

The steps can be summarized as below:

- Client sends a SYN packet to the server.
- Web server responds with SYN-ACK packet.

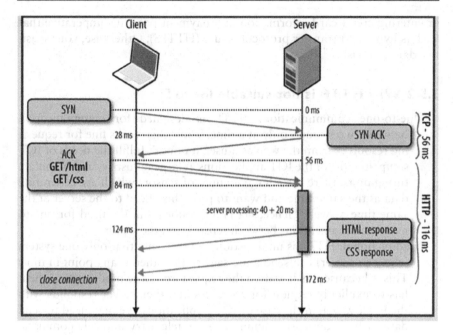

Figure 3.21 HTTP.

- Client again sends an ACK packet, concluding a connection establishment. This is also commonly referred to as a 3-way handshake.
- Client sends a HTTP request to the server asking for a resource.
- Client waits for the server to respond to the request.
- Web server processes the request, finds the resource and sends the response to client.
- If no more resources are required by the client, it sends a FIN packet to close the TCP connection.

3.5.1 HTTP protocol in IoT

Hyper Text Transfer Protocol (HTTP) is the most well-known example of an IoT network protocol. This protocol has formed the foundation of data communication over the web. It is the most common protocol which is used for IoT devices when there is a lot of data to be published. HTTP is the protocol used to transfer data over the web. HTTP uses a server–client model. A client might be a home computer, a laptop, or a mobile device. The HTTP server is typically a web host running web server software, such as Apache or IIS.

Is HTTP safe to use? The accurate answer is: it depends. If you are just browsing the web, looking at cat memes and dreaming about that $200 cable knit sweater, HTTP is fine. However, if you're logging into your bank or entering credit card information in a payment page, it's imperative that URL is hyper text transfer protocol secure (HTTPS). Otherwise, your sensitive data is at risk.

3.5.2 Why HTTP is not suitable for IoT?

One-to-one communication: HTTP is designed for communication between two systems only at a time. While this works fine for requesting resources from the web as a user, it doesn't fulfill the needs of IOT setup. In most of the IOT applications at large industries and manufacturing units, there are a large number of sensors which are generating data at the same time and want to push this ahead to the server at the same time as well. Hence, HTTP does not fulfill the need for one to many communications between sensors and the server.

Unidirectional: HTTP is unidirectional in the sense that only one system (client or server) can send a message to the other at any point in time. This is because it is based on the request–response model where client has to explicitly request for resources and then server responds with them. However, in the case of IOT applications, we may need to send data in both directions simultaneously telemetry and tele command can be executed at the same time!

Synchronous request–response: After requesting a resource to the server, the client has to wait for the server to respond. This blocks some system resources such as threads, CPU cycles etc. on both the client as well as the server side. Additionally, this leads to the slow transfer of data. IOT sensors are small devices with very limited computing resources and hence cannot work efficiently in a synchronous manner. All the widely used IOT protocols are based on asynchronous model.

Not designed for event-based communication: Most of the IOT applications are event-based. The sensor devices measure for some variables such as temperature, air quality and contents etc. and might need to take event-driven decisions like turning off a switch etc. HTTP was designed for a request–response-based communication rather than an event-driven communication. Also, programming this event-based system using HTTP protocol becomes a big challenge, especially because of the limited computing resources on the sensor devices.

Scalability: HTTP connections utilize high system resources, especially I/O threads. For every HTTP connection, the client/server also has to open an underlying persistent TCP connection. As more sensor devices are added to the network, the load on the server increases. If the sensor devices themselves are connected to multiple other devices, this puts a heavy load on the tiny system resources of the sensors. Hence, HTTP does not scale well for IoT applications.

High Power Consumption: Since HTTP utilizes heavy system resources as explained above, this also leads to heavy power consumption.

Advanced Wireless Sensor Networks of today have battery-operated sensor devices utilizing wireless network connections. Because of the heavy power consumption, HTTP is not suitable for advanced Wireless Sensor Networks.

As is clear from the above, HTTP has severe limitations for IoT applications. Many advanced application-layer protocols (MQTT, AMQP, CoAP) have been developed to overcome these limitations.

3.6 Wi-Fi

Wi-Fi is a wireless networking technology that allows devices such as computers (laptops and desktops), mobile devices (smartphones and wearables), and other equipment (printers and video cameras) to interface with the Internet as shown in Figure 3.22. It allows these devices and many more to exchange information with one another, creating a network. Wi-Fi has played a foundational role in delivering IoT innovation, providing pervasive connectivity to connect a wide variety of "things" to each other, to the Internet, and to the (at present) 18 billion Wi-Fi devices in use around the world. The economic potential of the Internet of Things is boundless, and Wi-Fi delivers a vast range of opportunities across a variety of sectors, including smart homes, smart cities, automotive, healthcare, enterprise, government, and IIoT environments.

Figure 3.22 Wi-Fi.

Wi-Fi enables users to automate their smart homes and connect a wide variety of connected household objects, to monitor supply chains and other critical functions in real time in industrial facilities, and to unlock business value by increasing productivity and efficiencies for both enterprises and hybrid-work scenarios. The integration and interoperability delivered by

Wi-Fi will enable IoT solutions to securely interconnect to one another and to billions of user-centric devices to unlock the greatest value from IoT applications and environments.

3.6.1 How does Wi-Fi work?

Wi-Fi is a wireless technology for networking, which uses electromagnetic waves to transmit networks. We know that there are many divisions of electromagnetic waves according to their frequency, such as X-rays, Gamma rays, radio waves, and microwaves. In Wi-Fi, the radio frequency is used. In transmitting the Wi-Fi signal, there are three mediums as shown in Figure 3.23:

- **Base station network or an Ethernet (802.3) connection:** This is the main host network from where the network connection is provided to the router.
- **Access point or router:** This is a bridge between a wired network and a wireless network. It accepts a wired Ethernet connection, converts the wired connection to a wireless connection and spreads the connection as a radio wave.
- **Accessing devices:** This is our mobile phone, computer, etc. from where we use the Wi-Fi and surf the Internet.

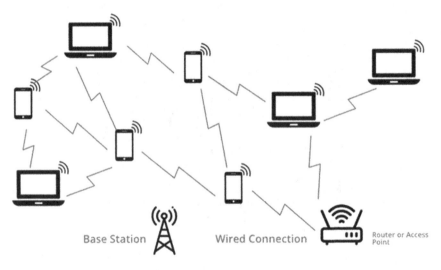

Figure 3.23 Working of Wi-Fi.

All the electronic devices read data in binary form, also router or our devices, here routers provide radio waves and those waves are received by our devices and read the waves in binary form. We all know how a wave looks like, the upper pick of the wave is known as 1 and the lower pick of the wave is known as 0 in binary, as in Figure 3.24.

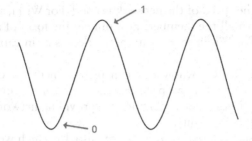

Figure 3.24 Data transmission.

3.6.2 Applications of Wi-Fi

Wi-Fi has many applications. It can be used in all the sectors where a computer or any digital media is used; it can also be used for the purposes of entertainment. Among the applications are the following:

- Accessing the Internet: Using Wi-Fi, we can access the Internet in any Wi-Fi-capable device wirelessly.
- We can stream or cast audio or video wirelessly on any device using Wi-Fi for our entertainment.
- We can share files, data, etc. between two or more computers or mobile phones using Wi-Fi, and the speed of the data transfer rate is also very high. Also, we can print any document using a Wi-Fi printer, this is very much used nowadays.
- We can use Wi-Fi as **Hotspots**, which creates Wireless Internet access over a particular range of area. Through the use of such hotspots, the owner of the main network connection can offer temporary network access to Wi-Fi-capable devices so that the users can use the network without knowing anything about the main network connection. Wi-Fi adapters are mainly spreading radio signals using the owner network connection to provide a hotspot.
- Using Wi-Fi or WLAN we can construct simple wireless connections from one point to another, known as point-to-point networks. This can be useful to connect two locations that are difficult to reach by wire, such as two buildings of corporate business.
- One more important application is **voice-over Wi-Fi (VoWi-Fi)**. Some years ago, telecom companies had introduced Voice over Long-Term Evolution(VoLTE). Nowadays they are introduced to VoWi-Fi, by which we can call anyone by using our home Wi-Fi network, only one thing is that the mobile needs to connect with the Wi-Fi. Then the voice is transferred using the Wi-Fi network instead of using the mobile SIM network, so the call quality is very good. Many mobile phones are already getting the support of VoWi-Fi.
- Wi-Fi in offices: In an office, all the computers are interconnected using Wi-Fi. In the case of Wi-Fi, there are no wiring complexities. In

addition, the speed of the network is good. For Wi-Fi, a project can be presented to all the members at a time in the form of an Excel sheet, PPT, etc. For Wi-Fi, there is no network loss as in cable due to cable break.

- Also using W-Fi a whole city can provide network connectivity by deploying routers at a specific area to access the internet. Already schools, colleges, and universities are providing networks using Wi-Fi because of its flexibility.
- Wi-Fi is used as a *positioning system* also, by which we can detect the positions of Wi-Fi hotspots to identify a device location.

3.7 TCP/UDP

Transmission Control Protocol (TCP) is a communications standard that software applications use to exchange data. It sets the parameters for the exchange, confirms what is being sent, where it is coming from, where it is going, and whether or not it arrived correctly. Unlike User Datagram Protocol (UDP), another standard that applications use to exchange data, TCP is designed for accuracy, rather than speed. In data transport, data packets can sometimes arrive out of order or be lost. TCP numbers each packet to ensure that every piece reaches its destination and can be rearranged if needed. When packets don't arrive within a specified timeframe, Transmission Control Protocol requests re-transmission of the lost data.

Throughout the entire exchange, TCP maintains the connection between the two applications, ensuring that both parties send and receive everything that needs to be transmitted and confirming that it's correct.

Transmission Control Protocol is the most popular standard for exchanging data over the Internet Protocol (IP), and it's often referred to as TCP/IP. Since it helps facilitate the exchange of data over the Internet, TCP is part of what's known as the "Transport Layer" of a network.

3.7.1 How Transmission Control Protocol works

Transmission Control Protocol (TCP) serves as an intermediary between two applications that need to exchange data. When an application wants to transmit data, TCP ensures that:

- Data arrives in order
- Data has minimal errors
- Duplicate data gets discarded
- Lost or discarded packets get resent

TCP act rather like a courier for the Internet. Once TCP establishes the connection and defines the interaction, it boxes up the data, loads it onto separate trucks, and sends it to its destination via the IP highway.

"On the road," traffic jams (network congestion), detours (traffic load balancing), and car accidents (network errors) can cause data to arrive out of order or prevent it from arriving on time.

As the data packets arrive, the recipient essentially signs for them with an acknowledgment: "I've received packet 4." If the sender does not receive this acknowledgement within a specified time, it resends the transmission. Once everything has arrived, TCP terminates the connection.

The entire process happens in three distinct phases, as detailed in the next section of the chapter.

3.7.2 The three phases of TCP operations

TCP operations involve numerous steps where the two endpoints use TCP to make requests, acknowledge each other, and confirm that the exchange is happening as intended. These steps fit into three main stages:

1. Connection establishment
2. Data transfer
3. Connection termination

The phase names are self-explanatory, but there's a lot more happening within each stage of the process.

During the **connection establishment phase**, TCP facilitates a "three-way handshake" where the applications request to synchronize and acknowledge each other. At this stage, TCP sets the parameters for the exchange and confirms that both entities (such as a server and a client) can participate in the exchange.

During the **data transfer phase**, TCP accepts the data being transferred, breaks it into ordered packets, adds a TCP header to provide context, and forwards it to the recipient using the Internet Protocol.

During the **connection termination phase**, the applications wait until they both acknowledge that the transmission is finished and error-free, and then TCP closes the connection between them.

3.7.3 TCP segments

When the Transmission Control Protocol receives a data stream, divides it up, and adds a TCP header to the transfer, the data stream becomes a "TCP segment." The TCP header ensures that when the individual data packets arrive at their destination, they can easily be arranged in the correct order, and the recipient can clearly see if anything is missing.

In the connection establishment phase, applications can announce their Maximum Segment Size (MSS), which defines the largest TCP segment they will exchange. This represents the largest amount of data that will be transmitted in a single segment. If the MSS is too large, IP fragmentation will break the individual packets into smaller pieces, increasing the risk that some packets will get lost and that the applications will have to retransmit the data multiple times (this increases latency).

3.7.4 Congestion control

During the data transmission TCP ensures that packets are sent at a pace that the network's resources can handle. Initially, TCP only allows a few bytes to go through the network. This is known as the "Congestion Window" (CWND). When the recipient acknowledges the data has arrived successfully, then TCP exponentially increases the CWND and allows more data to go through.

After the CWND reaches a specified threshold, the increase becomes linear. If a packet gets lost, TCP significantly reduces the Congestion Window and transmits slowly again. Over time, TCP data throughput forms a sawtooth pattern, where the transmission rate increases and decreases sharply to control network congestion.

3.7.5 Error detection

Unlike UDP, TCP checks transmissions for errors. Using sequence numbers and a checksum, it determines whether transmissions arrive correctly. If one bit is inaccurate, the checksum will be incorrect. When that happens, TCP drops the incorrect segment.

Basis	Transmission control protocol (TCP)	User datagram protocol (UDP)
Type of service	TCP is a connection-oriented protocol. Connection-orientation means that the communicating devices should establish a connection before transmitting data and should close the connection after transmitting the data.	UDP is the Datagram-oriented protocol. This is because there is no overhead for opening a connection, maintaining a connection, and terminating a connection. UDP is efficient for broadcast and multicast types of network transmission.
Reliability	TCP is reliable as it guarantees the delivery of data to the destination router.	The delivery of data to the destination cannot be guaranteed in UDP.
Error checking mechanism	TCP provides extensive error-checking mechanisms. It is because it provides flow control and acknowledgment of data.	UDP has only the basic error checking mechanism using checksums.

(Continued)

Basis	Transmission control protocol (TCP)	User datagram protocol (UDP)
Acknowledgment	Acknowledgment segment is present.	No acknowledgment segment.
Sequence	Sequencing of data is a feature of Transmission Control Protocol (TCP). This means that packets arrive in order at the receiver.	There is no sequencing of data in UDP. If the order is required, it has to be managed by the application layer.
Speed	TCP is comparatively slower than UDP.	UDP is faster, simpler, and more efficient than TCP.
Retransmission	Retransmission of lost packets is possible in TCP, but not in UDP.	There is no retransmission of lost packets in the User Datagram Protocol (UDP).
Header Length	TCP has a (20–60) bytes variable length header	UDP has an 8 bytes fixed-length header.

3.8 ADVANCED MESSAGE QUEUING PROTOCOL (AMQP)

Advanced Message Queuing Protocol (AMQP) is an open protocol for asynchronous message queuing, which has been developed and matured over several years as shown in Figure 3.25. AMQP is an open standard, binary application layer protocol designed for message-oriented middleware i.e., AMQP protocol standardizes messaging using producers, brokers, and consumers and messaging increases loose coupling and scalability.

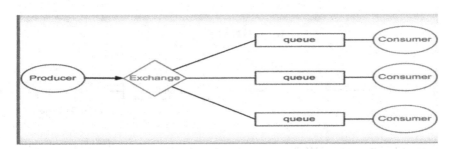

Figure 3.25 AMQP.

The AMQP was contrived to enable a wide range of applications and systems to work together, regardless of their internal designs, standardizing enterprise messaging on an industrial scale. AMQP protocol has been selected the OASIS industry standards group1, with the intention of it eventually becoming an ISO/IEC standard. The AMQP protocol is almost a complete superset, lacking only explicit protocol support for Last-Value-Queues and will messages. However, its deliberate design for extensibility, using an

IANA-like approach with a discursive approach, ensures that such features can be added in a forward-compatible, widely agreed-upon way.

3.8.1 How AMQP works

Let's see how AMQP works. It is a protocol that deals with both publishers and consumers. The publishers produce the messages, the consumers pick them up and process them. It's the job of the message broker (such as RabbitMQ) to ensure that the messages from a publisher go to the right consumers. In order to do that, the broker uses two key components:

1. Exchanges
2. Queues

The following Figure 3.26 shows how they connect a publisher to a consumer:

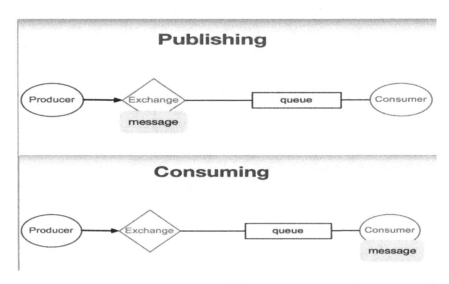

Figure 3.26 Working of AMQP.

As you can see, the setup is pretty straightforward. A publisher sends messages to a named exchange and a consumer pulls messages from a queue, or the queue pushes them to the consumer depending on the configuration. The connections have to be made in the first place, so the question is: how publishers do and consumers discover each other? The answer is via the name of the exchange. Usually, either the publisher or the consumer creates the exchange with a given name and then makes that name public. AMQP is designed to solve real problems completely.

3.9 CONSTRAINED APPLICATION PROTOCOL (CoAP)

IoT devices have very limited resources. For example, they have embedded processors or controllers, and limited RAM and ROM, and they need to operate on battery without replacement for weeks, months, or even years. Even they have to communicate data swiftly; although this may be only small in amounts, it was on limited network bandwidth. Bringing the web to constrained devices that lack the capabilities of computer and smartphones require a special type of IoT protocol. The Constraint Application Protocol (CoAP) is one such protocol designed to fit this requirement as shown in Figure 3.27.

CoAP is a specialized web transfer protocol for use with constrained nodes (low power sensors and actuators) and constrained networks (low power, lossy network). It enables those nodes to be able to talk with other constrained nodes over the Internet. The protocol is specifically designed for M2M applications, such as smart energy, home automation, and many industrial applications.

CoAP is a web-based protocol that has been specifically designed to connect small, semi-intelligent devices to the Internet of Things (IoT). The CoAP works with constrained nodes and constrained networks, to facilitate the compartmentalized deployment of machine-to-machine (M2M) solutions comprising of a multitude of network-enabled devices.

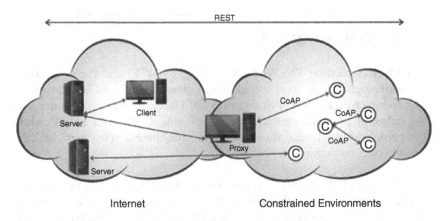

Figure 3.27 CoAP.

The CoAP protocol is necessary because traditional protocols such as TCP/IP are considered "too heavy" for IoT applications that involve constrained devices. CoAP protocol runs on devices that support UDP protocol. In UDP protocol, client and server communicate through connectionless data grams.

To put things a little more simply, CoAP Protocol facilitates the rapid networking of hundreds of IoT-enabled devices, to build a single networked application, such as could be used for automated manufacturing lines, or a smart building. CoAP is a specialized web transfer protocol for use with constrained devices (such as microcontrollers) and constrained networks in the IoT.

This protocol is used in M2M data exchange, such as smart energy, home automation, and many industrial applications and is very similar to HTTP. CoAP protocol is necessary because traditional protocols, such as TCP/IP, are considered too heavy for IoT applications that involve constrained devices. CoAP protocol runs on devices that support UDP protocol. In UDP protocol, client and server communicate through connectionless data grams as shown in Figure 3.28.

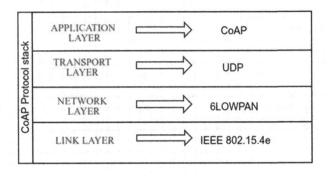

Figure 3.28 CoAP protocol stack.

CoAP Architecture: The World Wide Web (WWW) and the constraints ecosystem are the two foundational elements of the CoAP protocol architecture as shown in Figure 3.29. Here, the server monitors and helps in communication happening using CoAP and HTTP while proxy devices bridge the existing gap for these 2 ecosystem, making the communication smoother.

CoAP allows HTTP clients (also called CoAP clients here) to talk or exchange data/information with each other within resource constraints as shown in Figure 3.30. While one tries to understand this architecture, becoming familiar with some key terms is crucial:

- Endpoints are the nodes of which the host has knowledge.
- The client sends requests and replies to incoming requests.
- The server gets and forwards requests. It also gets and forwards the messages received in response to the requests it had processed.
- The sender creates and sends the original message.
- The recipient gets the information sent by the client or forwarded by the server.

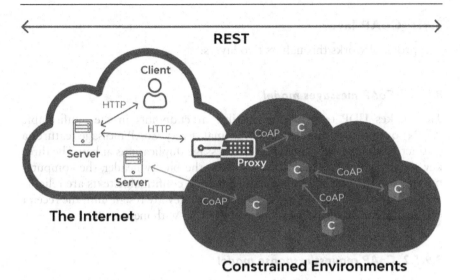

Figure 3.29 CoAP architecture.

Layer of TCP/IP	Layer in protocol stack	Layers in COAP protocol stack	
Application layer	Application layer	Request/response layer and message layer	CoAP
Transport layer	Transport layer	User datagram protocol (UDP)	
Network layer	Network layer	IPv6	
	Adaptation layer	6LoWPAN	
DataLink layer	DataLink layer	IEEE 802.15.4 MAC	
Physical layer	Physical layer	IEEE 802.15.4 PHY	

Figure 3.30 CoAP protocol stack.

3.9.1 CoAP layer

The protocol works through its two layers:

3.9.1.1 CoAP messages model

This makes UDP transactions possible at endpoints in the confirmable (CON) or non-confirmable (NON) format. Every CoAP message features a distinct ID to keep the possibilities of message duplications at bay. The three key parts involved to build this layer are the binary header, the computer option, and the payload. As explained before, confirmable texts are reliable and easy-to-construct messages that are fast; they are resent until the receipt of a confirmation of successful delivery (ACK) with message ID.

3.9.1.2 CoAP request/response model

This layer takes care of CON and NON message requests. Acceptance of these requests depend on the server's availability. Cases are:

1. If idle, the server will handle the request right away. If a CON, the client will get an ACK for it. If the ACK is shared as a token and differs from the ID, it is essential to map it properly by matching request-response pairs.
2. If there is a delay or a wait involved, the ACK is sent but as an empty text. When its turn arrives, the request is processed and the client gets a fresh CON.

The key traits of the request/response model are mentioned next:

- Request or response codes for CoAP are same as for the HTTP, except for the fact that they are in the binary format (0–8 byte Tokens) in CoAP's case.
- Request methods for making calls (GET, PUT, POST, and DELETE) are declared in the process.
- A CON response could either be stored in an ACK message or forward as CON/NON.

3.9.2 CoAP protocol security

The main concern from the security point of view is to provide data integrity, data authentication and data confidentiality. The CoAP provides security over Datagram Transportation Layer Security (DTLS) in the application layer as shown in Figure 3.31. As CoAP runs over the UDP protocol stack, there are chances of data loss or data disordering. But with DTLS security, these two problems can be solved. DTLS security adds three implementations to CoAP:

1. Packet retransmission
2. Assigning sequence number within handshake
3. Replay detection

The security is designed to prevent eavesdropping, tampering, or data forgery at any cost. Unlike network layer security protocols, DTLS in application layer protect end-to-end communication. DTLS also avoids cryptographic overhead problems that occur in lower layer security protocols.

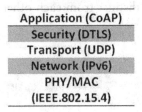

Figure 3.31 Image showing CoAP protocol at application layer in network architecture.

There is a Secured Handshake Mechanism in DTLS, as shown in Figure 3.32.

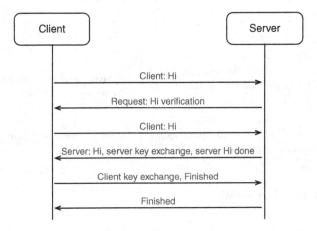

Figure 3.32 Image showing DTLS Secured Handshake Mechanism for CoAP.

The CoAP can also be implemented over TCP and over TLS. TCP and TLS transport for the Constrained Application Protocol (CoAP).

In one example of Client–Server Communication using the CoAP Protocol, an ESP8266 module is configured as a server and a browser on a laptop is configured as a client. The CoAP client sends some data to the server and the server acknowledges it by switching an LED on.

3.10 CoAP vs MQTT

As there are great similarities, we won't blame you if you consider these two identical. For instance, they both are used for IoT devices as they both necessitate less amounts of network packets causing more power-optimized performance, less storage consumption, and longer battery power.

CoAP and MQTT are distinct from each other in a number of different ways as shown in Figure 3.33.

CoAP vs MQTT

MQTT	CoAP
This model has publishers and subscribers as main participants	Uses requests and responses
Central broker handles message dispatching, following the optimal publisher to client path.	Message dispatching happens on a unicasting basis (one-to-one). The process is same as HTTP.
Event-oriented operations	Viable for state transfer
Establishing a continual and long-lasting TCP connection with the broker is essential for the client.	Involved parties use UDP packets (async) for message passing and communication.
No message labeling but have to use diverse messages for different purposes.	It defines messages properly and makes its discovery easy.

Figure 3.33 MQTT vs CoAP.

3.11 RF

Radio frequency (RF) is the oscillation rate of an alternating electric current or voltage or of a magnetic, electric, or electromagnetic field or mechanical system in the frequency range from around 20 kHz to around 300 GHz.

3.11.1 Electric current

Electric currents that oscillate at radio frequencies (**RF currents**) have special properties not shared by direct current or lower audio frequency alternating current, such as the 50 or 60 Hz current used in electrical power distribution.

- Energy from RF currents in conductors can radiate into space as electromagnetic waves (radio waves). This is the basis of radio technology.
- RF current does not penetrate deeply into electrical conductors but tends to flow along their surfaces; this is known as the skin effect.
- RF currents applied to the body often do not cause the painful sensation and muscular contraction of electric shock that lower-frequency currents produce. This is because the current changes direction too quickly to trigger depolarization of nerve membranes. However, this does not mean RF currents are harmless; they can cause internal injury as well as serious superficial burns called *RF burns*.
- RF current can easily ionize air, creating a conductive path through it. This property is exploited by "high-frequency" units used in electric arc welding, which use currents at higher frequencies than power distribution uses.
- Another property is the ability to appear to flow through paths that contain insulating material, like the dielectric insulator of a capacitor. This is because capacitive reactance in a circuit decreases with increasing frequency.
- In contrast, RF current can be blocked by a coil of wire, or even a single turn or bend in a wire. This is because the inductive reactance of a circuit increases with increasing frequency.
- When conducted by an ordinary electric cable, RF current has a tendency to reflect from discontinuities in the cable, such as connectors, and travel back down the cable toward the source, causing a condition called standing waves. RF current may be carried efficiently over transmission lines such as coaxial cables.

3.11.2 Applications of RF

- **Communications**: Radio frequencies are used in communication devices such as transmitters, receivers, computers, televisions, and mobile phones, to name just a few. Radiofrequencies are also applied in carrier current systems, including telephony and control circuits. The MOS integrated circuit is the technology behind the current proliferation of radio frequency wireless telecommunications devices such as cell phones.
- **Medicine**: Medical applications of radio frequency: Medical applications of radio frequency (RF) energy, in the form of electromagnetic waves (radio waves) or electrical currents, have existed for over 125

years and now include diathermy, hyperthermy treatment of cancer, electro surgery scalpels used to cut and cauterize in operations, and radiofrequency ablation. Magnetic resonance imaging (MRI) uses radiofrequency waves to generate images of the human body.

- **Measurement**: Test apparatus for radio frequencies can include standard instruments at the lower end of the range, but at higher frequencies, the test equipment becomes more specialized.

3.11.3 Frequency bands

Radio spectrum: The radio spectrum of frequencies is divided into bands with conventional names designated by the International Telecommunication Union (ITU) (Table 3.1).

Table 3.1 Frequency bands

Frequency range	Wavelength range	ITU designation		IEEE bands
		Full name	Abbreviation	
Below 3 Hz	$>10^5$km	Tremendously low frequency	TLF	—
3–30 Hz	10^5–10^4km	Extremely low frequency	ELF	—
30–300 Hz	10^4–10^3km	Super low frequency	SLF	—
300–3000 Hz	10^3–100km	Ultra low frequency	ULF	—
3–30kHz	100–10km	Very low frequency	VLF	—
30–300kHz	10–1km	Low frequency	LF	—
300kHz–3 MHz	1 km–100 m	Medium frequency	MF	—
3–30 MHz	100–10 m	High frequency	HF	HF
30–300 MHz	10–1 m	Very high frequency	VHF	VHF
300 MHz–3 GHz	1 m–100 mm	Ultra high frequency	UHF	UHF, L, S
3–30 GHz	100–10 mm	Super high frequency	SHF	S, C, X, Ku, K, Ka
30–300 GHz	10–1 mm	Extremely high frequency	EHF	Ka, V, W, mm
300 GHz–3 THz	1 mm–0.1 mm	Tremendously high frequency	THF	—

3.11.4 How is RF used?

The most important use for RF energy is in providing telecommunications services. Radio and television broadcasting, cellular telephones, personal communications services (PCS), pagers, cordless telephones, business radio, radio communications for police and fire departments, amateur radio, microwave point-to-point links and satellite communications are just a few of the many telecommunications applications of RF energy. Microwave ovens are an example of anon-telecommunication use of RF energy. Radiofrequency radiation, especially at microwave frequencies, can transfer energy to water molecules. High levels of microwave energy will generate heat in water-rich materials such as most foods. This efficient absorption of microwave energy via water molecules results in rapid heating throughout an object, thus allowing food to be cooked more quickly in a microwave oven than in a conventional oven. Other important non-telecommunication uses of RF energy include radar and industrial heating and sealing. Radar is a valuable tool used in many applications range from traffic speed enforcement to air traffic control and military surveillance. Industrial heaters and sealers generate intense levels of RF radiation that rapidly heats the material being processed in the same way that a microwave oven cooks food. These devices have many uses in industry, including molding plastic materials, gluing wood products, sealing items such as shoes and pocketbooks, and processing food products. There are also a number of medical applications of RF energy, such as diathermy and magnetic resonance imaging (MRI).

3.11.5 How are people exposed to RF radiation?

People can be exposed to RF radiation from both natural and human-made sources. The natural sources include:

- Outer space and the Sun
- The sky – including lightning strikes
- The earth itself – most radiation from the Earth is infrared, but a tiny fraction is RF Human-made RF radiation sources include:
- Broadcasting radio and television signals
- Transmitting signals from cordless telephones, cell phones and cell phone towers, satellite phones, and two-way radios
- Radar
- Wi-Fi, Bluetooth devices, and smart meters
- Some medical procedures, such as radio frequency ablation (using heat to destroy tumors)
- "Welding" pieces of polyvinyl chloride (PVC) using certain machines
- Millimeter wave scanners (a type of full body scanner used for security screening)

3.12 IPv4/IPv6

IP stands for Internet Protocol while v4 refers to the fourth version of the protocol (IPv4). In 1983, the ARPANET's major version, IPv4, was put in use for operation. The addresses it uses are 32-bit values that are written using decimal format.

3.12.1 The First major protocol

In the 1970s, the First major Internet Protocol was invented. This protocol was **IPv4 (Internet Protocol Version 4)**. This protocol was originally designed to be used as an isolated military network. Following its successful use in the military area, it was then also adapted for public use. The addresses used in IPv4 was 32 bit because in the 1970s, 32-bit was the biggest register found in any common processors. There were, however, a large number of limitations in **IPv4**, including:

- There were a large number of devices that were connected to the Internet throughout the world due to which there was a shortage of address spaces and the size of the address space was exhausting.
- Due to the insufficient size of the **IPv4**, It was not accommodating additional parameters, which were leading to weak protocol extensibility.
- Security was one of the major limitations Of **IPv4**. There was no limit to the information hosted on the network.
- Service support for **IPv4** was very poor.
- 50% of all addresses were reserved for the United States of America because that was the place where the Internet was born.
- Due to the increase in the number of servers connected to the Internet, there was also an increase in the number Of **IPv4** routers. These **IPv4** routers also started consuming addresses.

Due to the above-listed limitations, it was clear that a day would come when IPv4 address space would run out. IPv4 would not last forever. Therefore, in order to overcome the limitations, a better and new version of the Internet Protocol, **Internet Protocol Version 6 (IPv6)**, was constructed.

3.12.2 Components

Network: The part of the system which specifies the determinant assigned to it. The networking portion also specifies the connection classification which has been allocated.

Host part: The hosts include a system on a network in a special manner. Each client is given this portion of the IPv4 address. The node of every component of the system will be the same but the guest portion should differ.

Subnet number: The sub-network of IPv4 seems to be the subnet identifier. Regional systems with a large number of nodes are separated into subnets, each with a subnet identifier.

3.12.3 Benefits of IPv4

- Encrypted is possible with IPv4 protection to maintain safety and confidentiality.
- The IPv4 system allotment is considerable, with over 85,000 operational devices currently. It has become simple to connect several gadgets all over a wide network without using NAT.
- It is a networking paradigm that provides both good service and cost-effective knowledge dissemination.
- IPV4 addresses have been renamed to allow for perfect trans coding.
- Networking is much more flexible and cost-effective since naming is done more efficiently.
- In multichannel organizations, the transfer of data throughout the system is much more specialized.

3.12.4 Anatomy of IPv4 address

An IPv4 address actually consists of two parts: one that identifies your network and the other that identifies the host (i.e. the device) within the network. These parts aren't equal or fixed, so to determine the length of the network part, the address also has a "network mask." In CIDR notation, this is a number after a slash that determines how many bits of the address make up the network prefix. For example, 192.168.0.1/24 indicates that an IPv4 address 192.168.0.1 has a 24-bit long prefix and that the network it belongs to contains addresses ranging from 192.168.0.0 to192.168.0.255 (i.e. all having a common value of the first 24 bits). Thanks to routing protocols such as RIP, OSPF, and BGP, routers can inform each other about the network addresses that are assigned to them or that they "know of" from other routers, so that the data packets can be forwarded to the right network and therefore also the right device.

3.12.5 Dynamic IP addresses

When checking the IPv4 address of your own device, you may find out that it is not always the same. This is because the DHCP server will assign an IP address to your device dynamically, i.e., lease it for a specific amount of time. If your device doesn't request a DHCP lease renewal in time, then the IPv4 address will be released and assigned to a different device. This mechanism has been implemented to conserve the very limited pool of IPv4 addresses available. As a consequence of its architecture, the Internet Protocol version 4 is capable of providing 2^{32} or over 4 billion IP addresses (4 294 967 296, to be precise). If all of them were static we would only be able to provide roughly half our population with an IP-equipped device. Thanks to dynamic assignment, we've been able to manage with only IPv4 until just the last decade and have been actively using it alongside IPv6 since it was first introduced. But don't be fooled – IPv6 adoption is inevitable, and you will see more of that protocol as the years go by.

3.12.6 IPv4 limitations

Believe it or not, the IPv4 system in use today was developed by the United States Department of Defense back in the early 1980s. The protocol was originally deployed as part of the Advanced Research Projects Agency Network (ARPANET), which would later become the foundation for the operation of the modern Internet. The issue, however, is that IPv4 relies on 32-bit addresses, limiting the number of unique identifiers it can accommodate to around 4 billion, according to the Federal Communications Commission. This poses a major obstacle for large-scale IoT deployments moving forward, as there simply aren't enough IP addresses for all the devices planned for deployment. Many businesses have been able to get around this limitation by implementing additional layers of technology, such as network address translation, to allow multiple locally connected devices to share a single public IP address. But with a deluge of new M2M applications on the rise, this method will not remain sustainable for very much longer.

3.12.7 Introduction to IPv6

Internet Protocols are the set of rules which are used in addressing the packets of data so that they can travel across networks to arrive at the correct destination. Internet Protocol facilitates the exchange of data between two different computers. Data to be transferred through the Internet is divided up into small pieces, called packets. Each packet is recognized by IP information which also helps the routers send the packets to the right place. Internet Protocols are very useful and form the basis for the entire Web service. These Protocols are the medium between two different systems to connect with one another. Without these Protocols, using the Internet to transfer data between two devices would be impossible.

3.12.8 What is IPv6 in IoT?

Before going on IPv6, there might be a question in your mind that if IPv6 in IoT is the updated version of IPv4, then where is IPv5?

Well, the answer to your question is that the Internet Engineering Task Force (IETF) who built IPv4 decided to skip IPv5 as it would also eventually run out of addresses. Hence, they decided to directly jump on IPv6, where there will be no such concerns. IPv6 is the latest version of the Internet Protocol. Devices that use the Internet are recognized by their own IP addresses so that Internet communication can work. IPv6 in IoT identifies these devices so that they can be located through the Internet easily.

3.12.9 Advantages of IPv6 in the Internet of Things

- IoT is a vast field of technology. This field includes a large number of devices and their working is mainly focused via the Internet. IPv6 is capable of giving out various IP addresses to these IoT devices so that they can be easily recognized on the Internet and can work efficiently to transfer data from one IoT device to another.
- IPv6 networks have auto-configuration capabilities which are quite simple and can be managed easily in larger installations. With the help of this feature of IPv6, configuration effort and deployment costs in the field of IoT are reduced drastically.
- IPv6 is capable of sending large data packets simultaneously to conserve bandwidth with the help of the rapid transmission of data. Due to IPv6 in IoT, devices used in IoT will also be able to interact with each other.
- IPv6 provides far better security than IPv4. It also provides confidentiality, authenticity, and data integrity. This security given by IPv6 is of the utmost importance to IoT because of its high dependency on network.
- IPv6 in IoT has a highly efficient multicast communication feature that eliminates the requirement for routine broadcast messaging. This improvement helps in preserving the battery life of IoT devices by reducing the number of packets processed.
- IPv6 provides multiple addresses to devices. Its routing mechanism is also distributed in a better way than that of IPv4. With the help of this feature, programmers will have the liberty to assign IoT end-device addresses that are consistent with their own applications and network practices.

3.12.10 The IPv6 revolution

In contrast to IPv4, the IPv6 system is based on 128-bit addresses and is able to facilitate close to 340 undecillion unique IP identifiers. This is a massive increase in capability that promises to super-charge the IoT revolution, but that's not all the new system improves upon. IPv6 also supports auto-configuration, integrated security, and a variety of new mobility features, enabling a higher degree of network complexity. While the IPv6 system is not backwards-compatible with IPv4, both protocols are able to work in parallel without significant disruption. For example, upper layer protocols such as TCP and HTTP function in the same way for both systems.

The Internet Society, in partnership with several large companies, formally launched IPv6 in 2012 and has continued to campaign for its adoption over the past seven years. In 2018, the organization reported that around 25 percent of all Internet-connected networks possess IPv6 connectivity, with 49 countries delivering at least 5 percent of their traffic over the system.

According to Google, 24 of these countries have IPv6 traffic that exceeds 15 percent of total network usage, which is quite an accomplishment in such a short time frame. Broadband Internet service providers have been a driving force in the transition from IPv4, both in the U.S. and abroad. Comcast leads the charge in the U.S. with IPv6 deployment exceeding 66 percent, whereas British Sky Broadcasting has reached an impressive 86 percent in the U.K.

3.12.11 What are the risks?

DDoS and phishing attacks, data theft, and remote hacking of industrial control systems, healthcare systems and automotive technologies are all likely to carry over from IPv4 to IPv6. Mindlin explained that although the structural elements of IPv6 are naturally beefed up, along with the huge address space, hackers will eventually find the network vulnerabilities and, after that, get to work on attacking the higher layers.

It is said that with a vast global amount of reachable IPv6 addresses, each connected device can be connected directly to the internet network, increasing its visibility and also potentially emphasizing its vulnerability issues.

3.12.12 The role of automation

Like numerous business processes both within and outside of the IT space, IPv6 implementation can be made easier with the help of automation. Williams said while many tool sets are IPv6-aware, the fundamental lack of education around IPv6 means they may not be consistently used.

As it is with threat modelling, the relative immaturity of the IPv6 space means there is a dearth of tools currently available.

3.13 6LoWPAN

6LoWPAN (IPv6 over Low-Power Wireless Personal Area Networks), is a low-power wireless mesh network where every node has its own IPv6 address. This allows the node to connect directly with the Internet using open standards.

6LoWPAN came to exist from the idea that the Internet Protocol could and should be applied even to the smallest devices, and that low-power devices with limited processing capabilities should be able to participate in the Internet of Things.

3.13.1 Advantages of 6LoWPAN

Uses Open IP Standards

Offers End-To-End IP addressable Nodes

Offers Self-Healing, Robust and Scalable Mesh Routing

Leaf Nodes Can Sleep For a Long Duration of Time

Offers Thorough Support For The PHY Layer

It is a Standard: RFC 6282

- It works great with open IP standard including TCP, UDP, HTTP, COAP, MATT and web-sockets.
- It offers end-to-end IP addressable nodes. There's no need for a gateway; only a router which can connect the 6LoWPAN network to IP.
- It supports self-healing, robust, and scalable mesh routing.
- Offers both one-to-many & many-to-one routing.
- The 6LoWPAN mesh routers can route data to other nodes in the network.
- In a 6LowPAN network, leaf nodes can sleep for a long duration of time.
- It also offers thorough support for the PHY layer which gives freedom of frequency band & physical layer, which can be used across multiple communication platforms like Ethernet, Wi-Fi, 802.15.4 or Sub-1GHz ISM with interoperability at the IP level as shown in Figure 3.34.
- It is a standard: RFC6282.

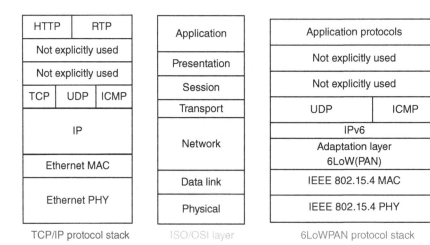

Figure 3.34 6LoWPAN protocol stack.

3.13.2 6LoWPAN application areas

With many low-power wireless sensor networks and other forms of wireless networks designed to tackle specific problems, it is essential that any new wireless system has a defined area which it addresses. While there are many forms of wireless networks, including wireless sensor networks, 6LoWPAN addresses an area that is currently not addressed by any other system, for example, that of using IP, and in particular IPv6 to carry the data.

The overall system is aimed at providing wireless internet connectivity at low data rates and with a low duty cycle. However, there are many applications where 6LoWPAN is being used:

- **Automation:** There are enormous opportunities for 6LoWPAN to be used in many different areas of automation.
- **Industrial monitoring:** Industrial plants and automated factories provide a great opportunity for 6LoWPAN. Major savings can be made by using automation in everyday practices. Additionally, 6LoWPAN can connect to the cloud which opens up many different areas for data monitoring and analysis.
- **Smart Grid:** Smart grids enable smart meters and other devices to build a micro-mesh network. They are able to send data back to the grid operator's monitoring and billing system using the IPv6.
- **Smart Home:** By connecting your home IoT devices using IPv6, it is possible to gain distinct advantages over other IoT systems.

3.13.3 6LoWPAN security

It is anticipated that the IoT will offer hackers a huge opportunity to take control of poorly secured devices and also use them to help attack other networks and devices.

Accordingly, security is a major issue for any standard such as 6LoWPAN, and it uses AES-128 link layer security which is defined in IEEE 802.15.4. This provides link authentication and encryption.

Further security is provided by the transport layer security mechanisms that are also included. This is defined in RFC 5246 and runs over TCP. For systems where UDP is used the transport layer protocol defined under RFC 6347 can be used, although this may require some specific hardware requirements.

3.13.4 6LoWPAN interoperability

One key issue of any standard is that of interoperability. It is vital that equipment from different manufacturers can operate together.

When testing for interoperability, it is necessary to ensure that all layers of the OSI stack are compatible. To ensure that this can be achieved, there are several different specifications that are applicable.

Any item can be checked to conform it meets the standard, and also directly tested for interoperability. 6LoWPAN is a wireless/IoT-style standard that has quietly gained significant ground. Although initially aimed at usage with IEEE 802.15.4, it is equally able to operate with other wireless standards, making it an ideal choice for many applications.

6LoWPAN uses IPv6 and this alone has given it a distinct advantage over other systems. With the world migrating towards IPv6 packet data, a system such as 6LoWPAN offers many advantages for low-power wireless sensor networks and other forms of low-power wireless networks.

Basic Requirements of 6LoWPAN:

- The device should have a sleep mode in order to support battery saving
- Minimal memory requirement
- Routing overheads should be lowered

Features of 6LoWPAN:

- It is used with IEEE 802.15,.4 in the 2.4 GHz band. Outdoor range: ~200 m (maximum)
- Data rate: 200kbps (maximum)
- Maximum number of nodes: ~100

Advantages of 6LoWPAN:

- 6LoWPAN is a mesh network that is robust, scalable, and can heal on its own. It delivers low-cost and secure communication in IoT devices.
- It uses IPv6 protocol and so it can be directly routed to cloud platforms. It offers both one-to-many and many-to-one routing.
- In the network, leaf nodes can be in sleep mode for a longer duration of time.

Disadvantages of 6LoWPAN:

- It is comparatively less secure than Zigbee.
- It has a lower immunity to interference than either Wi-Fi or Bluetooth. Without the mesh topology, it supports only a short range.

Chapter 4

Introduction to sensors and actuators

The new data economy greatly benefits from the Internet of Things (IoT). An IoT system's value extends beyond its initial intended use case, such as in automation. This is due to the fact that an IoT system's intelligence has additional value. The source of IoT data is a sensor. Additionally, IoT sensors and actuators can cooperate to provide automation on an industrial scale. Finally, over time, analysis of the data generated by these sensors and actuators might yield insightful business information.

Sensor technology is evolving at a rate never before witnessed, driven by advances in materials science and nanotechnology. As a result, it is becoming more accurate, smaller, and less expensive, and able to measure or detect things that weren't previously imaginable. In fact, in a few years we'll see a trillion new sensors deployed annually since sensing technology is advancing so quickly.

4.1 INTRODUCTION TO SENSORS

A sensor might be more accurately called a transducer. This term can be applied to any physical object that transforms one form of energy into another. In a sensor, a physical phenomenon is transformed into an electrical impulse by the transducer, which then decides the reading. A microphone, for example, is a sensor that converts sound waves from vibration energy to electrical energy in a usable fashion so that other parts of the system may link the electrical energy back to the original sound. A transducer is a general phrase that refers to both sensors and actuators. A transducer changes a signal from one kind of energy into another.

For instance, sound waves are transformed into electrical signals in a microphone for amplification before being transferred to an output device like a loudspeaker. This serves as an illustration of a transducer as shown in Figure 4.1.

DOI: 10.1201/9781003307488-4

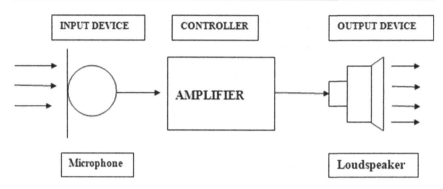

Figure 4.1 Transducer.

A sensor is a device that generates an output signal in order to detect physical phenomena. By transforming them into another form, usually electrical pulses, it detects, measures, or indicates any specific physical quantity such as light, heat, motion, pressure, or similar phenomena. You can set up sensors so that they react when specific changes take place. For instance, you could programme the sensor to notify an operator if a room's temperature rises too high. In order to warn the control center, the sensor converts the heat's physical input into an electrical signal.

4.2 INTRODUCTION TO ACTUATORS

An actuator is a different kind of transducer which is used in many IoT systems. Simply put, an actuator works the opposite way from a sensor, transforming an electrical input into physical movement. The actuators are in motion. In other words, based on what has been detected, they take certain physical acts. Actuators come in a variety of forms, such as electric motors, hydraulic systems, and pneumatic systems. A shutdown valve is a nice illustration of an actuator. It closes the valve when it gets a signal from a sensor or control module. An electrical signal is sent into the actuator, which converts it into a physical action.

4.3 CONTROLLER

A sensor may gather data and send it to a control center in a general IoT setup as shown in Figure 4.2. There, the judgment is governed by previously established rationale. Therefore, in response to the sensed input, a corresponding command operates an actuator. As a result, with the IoT, sensors and actuators collaborate from opposing ends. We'll talk more about the location of the control center within the larger IoT system later.

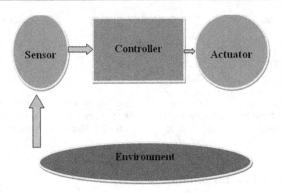

Figure 4.2 Working of controller.

Simply put, the IoT is a collection of numerous deeply connected technologies rather than a single technology. The gathering, handling, communication, and analysis of data present numerous difficulties. These IoT devices gather a lot of data, and it is up to the user to choose which information is pertinent for their circumstance, where to process or store it, and the preferred level of communication. Data can be stored, preprocessed, and processed on a remote server or right at the network's edge.

4.4 WHAT CONNECTS SENSORS AND ACTUATORS IN IoT DEVICES?

The sensor gathers data and transmits it to the control center in an intelligent IoT system. According to its programming, the control center processes the data before giving directions to the actuators to carry out certain actions. The basic distinction between sensors and actuators in the Internet of Things can be summarized as follows: the sensor is the brain, and the actuator is the limb that executes the tasks. Let's now develop that model further.

The fundamental infrastructure of an IoT framework is composed of sensors, actuators, computation servers, and the communication network as shown in Figure 4.3. Middleware is one of the technologies that is occasionally required. Software known as middleware serves as a conduit between an operating system, database, and applications, particularly those running over a network. All autonomous IoT components can be managed and connected via middleware.

Figure 4.3 Sensor to actuator flow.

4.4.1 Sensors characteristics

1. Static
2. Dynamic

4.4.2 Static characteristics

This refers to how a sensor's output alters in response to an input change after reaching a steady state.

- **Accuracy**: Accuracy is the capacity of measuring devices to produce a result that is reasonably close to the actual value of the quantity being measured. It counts mistakes. It is measured by absolute and relative errors. Compare the output's accuracy to a more advanced earlier system.

    Absolute error = Measured value – True value

    Relative error = Measured value/True value
- **Range**: This reveals the physical quantity's greatest and lowest values that the sensor is truly capable of sensing. There is no sense or form of reaction outside of these values.

    For instance, the temperature measuring range of a Resistive Temperature Device (RTD) is between −200°C and 800°C.
- **Resolution**: Resolution is a crucial parameter when choosing a sensor. The precision improves with increased resolution. Threshold is the condition where the accretion is equal to zero. It gives the slightest adjustments to the input that a sensor is capable of detecting.
- **Precision**: When repeatedly measuring the same quantity under the same set of guidelines, a measuring instrument must be able to produce the same reading. It implies agreement between successive readings, NOT closeness to the true value.

It has to do with the variability of a collection of measurements. It is a prerequisite for accuracy, but is not a sufficient condition.

- **Sensitivity**: Sensitivity describes the ratio of small changes in the system's response to small changes in its input parameters. It can be determined from the slope of a sensor's output characteristics curve. The little quantity difference will cause the instrument's reading to change.
- **Linearity**: The sensor value curve's departure from a specific straight line. The calibration curve determines the linearity. Under static circumstances, the static calibration curve plots the output amplitude versus the input amplitude. The linearity of a curve is expressed by its slope, which resembles a straight line.
- **Drift**: The variation in the sensor's measurement from a particular reading when kept at that value for an extended length of time.
- **Repeatability**: The variation in measurements made sequentially under the same circumstances. The measurements must be taken over a brief enough period of time to prevent considerable long-term drift.

4.4.3 Dynamic characteristics

In this, there are a number of properties of the systems, such as:

Zero-order system

- The output displays a prompt response to the input signal. Energy is not included for storing elements.
- E.g. potentiometer measure, linear and rotary displacements.

First-order system

- The output steadily approaches its ultimate value. It consists of a component for storing and dissipating energy.

Second-order system

- Complex output response makes up it. Before reaching steady state, the sensor's output response oscillates.

4.5 SENSOR CLASSIFICATION

- Passive & Active
- Analog & digital
- Scalar & vector

    1. **Passive Sensor**: It measures the amount of sunlight that is reflected back from the Sun as shown in Figure 4.4. It is unable to sense the

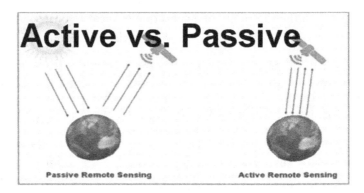

Figure 4.4 Active and passive sensor.

input on its own. Example: Accelerometer, soil moisture, water level and temperature sensors. Sleep, stress, light-dependent resistor (LDR), cameras are passive sensors with the flash turned off.

2. **Active Sensor**: These are those sensors that sense the input independently. It has a separate light or illumination source. Radar is an example. When the flash is activated, cameras function as active sensors. It is utilized to keep track of industrial machines in manufacturing and networking environments.

3. **Analog Sensor**: The response or output of the sensor is some continuous function of its input parameter as shown in Figure 4.5. Ex: Temperature sensor, LDR, analog pressure sensor, and analog hall effect.

4. **Digital sensor**: They respond in a binary fashion. They are designed to overcome the disadvantages of analog sensors. Along with the analog sensor, it also comprises extra electronics for bit conversion. Example: Passive infrared (PIR) sensor and digital temperature sensor (DS1620).

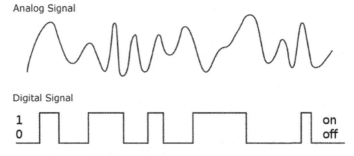

Figure 4.5 Analog and digital sensor.

5. **Scalar sensor:** It detects the input parameter only based on its magnitude. The answer for the sensor is a function of magnitude of some input parameter. They are not affected by the direction of input parameters. Example: temperature, gas, strain, color, and smoke sensor.

6. **Vector sensor:** The response of the sensor depends on the magnitude of the direction and orientation of input parameter. Example: Accelerometer, gyroscope, magnetic field and motion detector sensors.

4.6 IoT SENSOR TYPES

Sensors are made to react to a certain range of physical situations. They then produce a signal (often electrical) that might reflect the severity of the condition being measured. Light, heat, sound, distance, pressure, or another more particular circumstance, such as the presence or absence of a gas or liquid, may be among those conditions. The usual IoT sensors that will be used are as follows:

- Temperature sensors
- Pressure sensors
- Motion sensors
- Level sensors
- Image sensors
- Proximity sensors
- Water quality sensors
- Chemical sensors
- Gas sensors
- Smoke sensors
- Infrared (IR) sensors
- Ultrasonic sensors
- Acceleration sensors
- Gyroscopic sensors
- Humidity sensors
- Optical sensors

A description of each of these sensors is provided below in Figure 4.6.

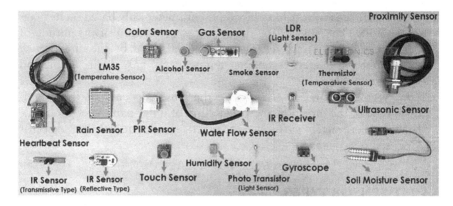

Figure 4.6 Different types of sensors.

4.6.1 Temperature sensors

Temperature sensors detect the temperature of the air or a physical object and convert that temperature level into an electrical signal that can be calibrated to accurately reflect the measured temperature as shown in Figure 4.7. These sensors could monitor the temperature of the soil to help with agricultural output or the temperature of a bearing operating in a critical piece of equipment to sense when it might be overheating or nearing the point of failure.

Example: Air conditioners, refrigerators, manufacturing processes, agriculture, and the healthcare industry.

Types of temperature sensors:

- ICs (like LM35)
- Thermistors
- Thermocouples
- RTD (Resistive Temperature Devices) etc.

Figure 4.7 Temperature sensor DS1820.

The most commonly used temperature sensor is DHT11 as shown in Figure 4.8.

Figure 4.8 DHT11.

4.6.2 Pressure sensors

These measure the pressure or force per unit area applied to the sensor and can detect things such as atmospheric pressure, the pressure of a stored gas or liquid in a sealed system such as tank or pressure vessel, or the weight of an object as shown in Figure 4.9.

Figure 4.9 BMP 180.

BMP180: a popular digital pressure sensor for use in mobile phones, PDAs, GPS navigation devices, and outdoor equipment.

4.6.3 Motion sensors

The movement of a physical object can be detected by motion sensors or detectors utilizing a variety of technologies, such as passive infrared (PIR), microwave detection, or ultrasonic, which detects objects using sound. These sensors can automate the control of doors, sinks, air conditioning and heating, and other systems in addition to being employed in security and intruder detection systems. A motion sensor, sometimes known as a motion detector, is a device used to detect and record movement as shown in Figure 4.10. In addition to phones, paper towel dispensers, game consoles, and virtual reality headsets, motion sensors are commonly employed in home and commercial security systems. Motion sensors are often embedded systems that consist of three main parts: a sensor unit, an embedded computer, and hardware, unlike many other types of sensors (which can be handled and isolated). Because motion sensors can be configured to carry out incredibly specialized tasks, these three components come in a variety of sizes and configurations. Motion sensors, for instance, can be used to turn on floodlights, set off audio alarms, turn on switches, and even call the police.

Motion sensors come in two varieties: active motion sensors and passive motion sensors. A transmitter and a receiver are both present in active sensors. This kind of sensor measures variations in the amount of sound or radiation that is reflected back into the receiver in order to detect motion. An electric pulse is sent to the embedded computer when something disrupts or modifies the sensor's field, and the embedded computer then communicates with the mechanical part. Ultrasonic sensor technology is used by the most popular kind of active motion detectors; these motion sensors produce sound waves to detect the presence of things. Additionally, tomographic sensors and microwave sensors (which produce microwave radiation) exist (which transmit and receive radio waves).

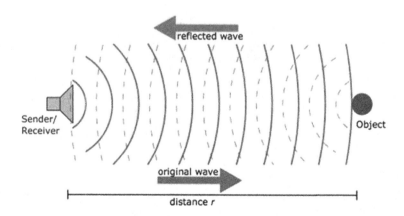

Figure 4.10 Active motion sensor.

In contrast to an active motion sensor, a passive motion sensor has no transmitter. The sensor detects motion based on a perceived rise in radiation in its environment rather than recording a steady reflection. The passive infrared (PIR) sensor is the most commonly used passive motion sensor type in home security systems. The PIR sensor is designed to detect the infrared radiation that the human body naturally emits. Only infrared is permitted to pass through the filter that houses the receiver, with a positive charge being produced in the receiver when a human enters the PIR sensor's field of detection. The sensing unit responds to this perceived change by sending electrical data to the hardware component and embedded computer.

4.6.4 Level sensors

A level sensor is a device used to maintain, measure, and monitor liquid (and occasionally solid) levels as shown in Figure 4.11. The sensor turns the observed data into an electric signal after detecting the liquid level. Although they can be found in many home products, such as ice makers in refrigerators, level sensors are most frequently utilized in the manufacturing and automotive industries. Level sensors can be divided into two categories: point level sensors and

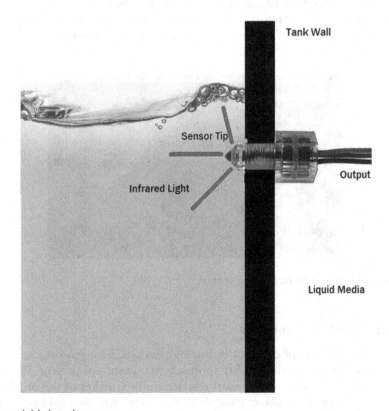

Figure 4.11 Level sensors.

continuous level sensors. In order to show whether a liquid has entered a certain point in a container, point level sensors are used. On the other side, precise liquid level measurements are provided by continuous level sensors. Invasive and non-contact level sensors are further categories of level sensors. Non-contact sensors use sound or microwaves, whereas invasive sensors establish physical contact with the object they are monitoring.

4.6.5 Image sensors

Images are captured by image sensors and saved digitally for processing as shown in Figure 4.12. Two of the most prominent examples of the use of this technology are facial recognition software and licence plate readers. Image sensors can be used in automated production lines to identify quality problems, such as how effectively a surface is painted after leaving the spray booth. Digital cameras, medical imaging systems, night-vision equipment, thermal imaging equipment, radars, sonars, media houses, and biometric systems all contain these sensors.

- Through an IoT network, these sensors are employed in the retail sector to keep track of customers entering the store.
- Through IoT networks, they are employed in offices and corporate buildings to keep an eye on workers and varied activities.

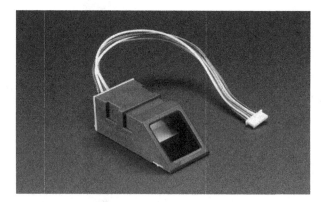

Figure 4.12 Finger print imaging sensor.

4.6.6 Proximity sensors

Through a number of different technology designs, such sensors can determine whether or not things that approach the sensor are present. Without any physical contact, these sensors can determine whether or not an object is nearby. In order to detect changes in the electromagnetic field or return signal, a proximity sensor frequently emits an electromagnetic field or a beam of electromagnetic radiation (infrared, for example). These are mostly employed in process control, monitoring, and object counting.

Example: A cell phone (comprised of an infrared LED and an IR light detector). A proximity sensor detects how close the phone is to an outside object, such as your ear. This sensing is done to reduce display power consumption while you're on a call by turning off the LCD backlight.

These approaches include:

Inductive technologies, which are useful for the detection of metal objects. These are contactless sensors used to only detect metal objects. Common applications:

- Industrial usages
- Production automation machines that count products and product transfers
- Security usages
- Detection of metal objects, armory, land mines, etc.

Capacitive technologies, which function on the basis of objects having a different dielectric constant than that of air. These are contactless sensors that detect both metallic and non-metallic objects, including liquid, powders, and granular. Common applications:

- Industrial usages
- Production automation machines that count products, product transfers
- Filling processes, pipelines, inks, etc.
- Fluid level, composition, and pressure
- Moisture control
- Non-invasive content detection
- Touch applications
- Photoelectric technologies, which rely on a beam of light to illuminate and reflect back from an object, or ultrasonic technologies, which use a sound signal to detect an object nearing the sensor detecting the presence of objects through emitting a high-frequency ultrasonic range

Common applications

- Distance measurement
- Anemometers for wind speed and direction detection
- Automation production processes
- Fluid detection
- Robotics

4.6.6.1 Water quality sensors

The requirement to be able to feel and quantify characteristics related to water quality is dictated by the significance of water to humans on earth, not only as a source of drinking water but also as a vital component in many industry processes. Some examples of what is sensed and monitored include:

- Chemical presence (such as chlorine levels or fluoride levels)
- Oxygen levels (which may impact the growth of algae and bacteria)
- Electrical conductivity (which can indicate the level of ions present in water)
- pH level (a reflection of the relative acidity or alkalinity of the water)
- Turbidity levels (a measurement of the amount of suspended solids in water)

4.6.7 Chemical sensors

Chemical sensors can be used to monitor the conditions of industrial processes by identifying the presence of specific chemicals that may have unintentionally seeped from their containers into areas where people are present.

Chemical sensors come in a wide variety of forms and are all purpose-built for particular tasks, but they all have two things in common: receptors and transducers. The receptor is the area of the chemical sensor where the analyte actually makes contact with it. The receptor's interactions with the analyte vary depending on the sensor. For instance, certain receptors may separate out particular molecules whereas others can react chemically with the analyte as a whole. The latter are referred to as more selective (sensors that target molecules in an analyte).

The transducer is the second element that all chemical sensors have in common. Transducers are in charge of absorbing the chemical data from the interaction of the receptor and analyte and transforming it into the appropriate electrical data. Then, a mechanical part or computer receives this information. The transducer may change the resistance, cause an audible alarm, or display the information on a screen (interface).

4.6.8 Gas sensors

Gas sensors, which are similar to chemical sensors, are calibrated to detect flammable, poisonous, or combustible gas nearby as shown in Figure 4.13. Depending on how precisely you wish to detect gas, you can change the sensitivity of the smoke sensor's built-in potentiometer Different types of gas sensors are shown in Table 4.1.

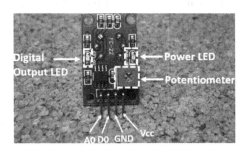

Figure 4.13 Gas sensor.

Table 4.1 Gas sensors

Sensor name	Gas to measure
MQ-2	Methane, Butane, LPG, Smoke
MQ-3	Alcohol, Ethanol, Smoke
MQ-4	Methane, CNG Gas
MQ-5	Natural gas, LPG
MQ-6	LPG, butane

How does it Work?

- The voltage that the sensor outputs changes accordingly to the smoke/gas level that exists in the atmosphere. In other words, the relationship between voltage and gas concentration is the following and is also shown in Figure 4.14:
- The greater the gas concentration, the greater the output voltage
- The lower the gas concentration, the lower the output voltage

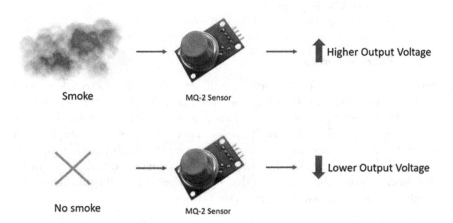

Figure 4.14 Gas sensor working.

Uses:

- In industries to monitor the concentration of the toxic gases
- In households to detect an emergency incident
- Used at hotels to avoid customers from smoking
- Used in air quality check at offices
- Used in air conditioners to monitor CO_2 levels
- Used in detecting fire
- Used to check the concentration of gases in mines

Examples of specific gases that can be detected include:

- Bromine (Br_2)
- Carbon Monoxide (CO)
- Chlorine (Cl_2)
- Chlorine Dioxide (ClO_2)
- Ethylene (C_2H_4)
- Ethylene Oxide (C_2H_4O)
- Formaldehyde (HCHO)
- Hydrazine(s)
- Hydrogen (H_2)
- Hydrogen Chloride HCl)
- Hydrogen Cyanide (HCN)
- Hydrogen Peroxide (H_2O_2)
- Hydrogen Sulfide (H_2S)
- Nitric Oxide (NO)
- Nitrogen Dioxide (NO_2)
- Ozone (O_3)
- Peracetic Acid ($C_2H_4O_3$)
- Propylene Oxide (C_3H_6O)
- Sulfur Dioxide (SO_2)

4.6.9 Smoke sensors

Smoke sensors or detectors use optical sensors (photoelectric detection) or ionization detection to detect the presence of smoke conditions, which could be an indicator of a fireas shown in Figure 4.15. Smoke detectors have long been a fixture in both homes and workplaces. Their application has been easier to use and more convenient with the development of the IoT. Additionally, giving smoke detectors a wireless connection makes it possible for them to have other functions that improve security and convenience Air quality sensor is shown in Figure 4.16.

Figure 4.15 MQ2 smoke sensor.

Figure 4.16 Air quality gas sensor.

4.6.10 Infrared (IR) sensors

Technologies using infrared (IR) sensors pick up the infrared light that is released by these objects. These kinds of sensors are used by non-contact thermometers to measure an object's temperature without having to touch it directly with a probe or sensor as shown in Figure 4.17. They find use in analyzing the heat signature of electronics and detecting blood flow or blood pressure in patients.

PIR Sensor, also known as a PIR(motion) sensor or IR sensor, is an abbreviation for passive infrared sensor, which is used to detect human or particle movement within a specific range. It has been widely embraced by the

Figure 4.17 IR sensor.

open-source hardware community for projects including the Arduino and Raspberry Pi due to its robust function and low cost advantages. It is frequently utilized in applications for automatic lighting and security alarms.

- The module actually consists of a pyroelectric sensor which generates energy when exposed to heat as shown in Figure 4.18.
- When a human or animal body will get in the range of the sensor it will detect a movement because the human or animal body emits heat energy in a form of infrared radiation.
- That's where the name of the sensor comes from, a passive infrared (PIR) sensor. The term "passive" means that the sensor is not using any energy for detecting purposes; it simply works by detecting the energy given off by the other objects.

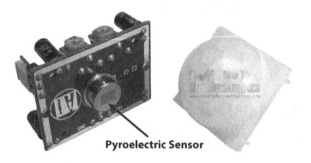

Pyroelectric Sensor

Figure 4.18 Pyroelectric sensor.

- PIR sensors sense general movement, but don't have information on who moved or what as shown in Figure 4.19 and Figure 4.20. An active IR sensor is necessary for this purpose.
- It does not emit the referred IR signals itself; rather, it passively detects the infrared radiations coming from the human body in the surrounding area.
- The detected radiations are converted into an electrical charge, which is proportional to the detected level of the radiation.
- Used in dark and light time.

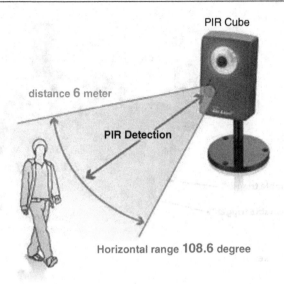

Figure 4.19 PIR Sensor.

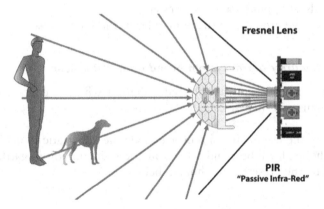

Figure 4.20 PIR Sensor.

- The module has just three pins, a Ground and a VCC for powering the module and an output pin which gives high logic level if an object is detected as shown in Figure 4.21.
- It has two potentiometers. One for adjusting the sensitivity of the sensor and the other for adjusting the time the output signal stays high when object is detected. This time can be adjusted from 0.3 seconds up to 5 minutes.
- Second is distance adjust.

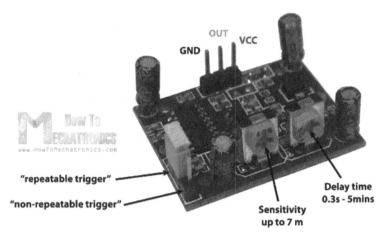

Figure 4.21 PIR sensor.

4.6.10.1 PIR sensor-based automatic door opening system

- Hotels, shopping malls, theatres etc.
- Senses the presence of the human body and sends pulses

4.6.10.2 Security alarm system based on a PIR sensor

- This sensor senses the infrared radiation which is emitted from the humans and then gives a digital output

Example of the use of such a sensor include the automatic switching on of outdoor lights, lift lobby, and the automatic switching on of garden lights, triggered by the presence of a human being.

4.6.10.3 What does a PIR sensor detect?

- The detector itself does not emit any energy but passively receives it, detecting infrared radiation from the environment. Once there is infra-red radiation from the human body/particle with temperature, focusing on the optical system causes the pyro electric device to generate a sudden electrical signal and an alarm is issued.

4.6.10.4 What is the difference between the PIR sensor and the motion detector sensor?

- An electronic device used to detect the **physical movement** (motion) in a given area and it transforms motion into an electric signal, the motion of any object or the motion of human beings.

- Used in the security industry. Businesses utilize these sensors in areas where no movement should be detected at all times, and it is easy to notice anybody's presence with these sensors installed.
- Used for intrusion detection systems, automatic door control, toll plaza, automated sinks, dryers, automated lighting, air conditioning, fan.
- PIR is only one of the technical methods to detect motion, so we will say PIR sensor is a subset of motion sensor. Because of PIR sensor are small in size, cheap in price, low-power and very easy to understand.

4.6.11 Ultrasonic sensors

An ultrasonic sensor is a device that uses ultrasonic sound waves to measure a target object's distance and then turns the sound that is reflected back into an electrical signal as shown in Figure 4.22. The transmitter (which generates sound using piezoelectric crystals) and the receiver are the two major parts of an ultrasonic sensor (which encounters the sound after it has travelled to and from the target). The sensor measures the amount of time that passes between the transmitter's sound emission and its contact with the receiver in order to determine the distance between the object and the sensor. $D = \frac{1}{2} T \times C$ (where D is the distance, T is the time, and C is the speed of sound ~ 343 meters/second) is the formula for this computation. When an object or obstruction gets in its way, the ultrasonic it generates at a frequency of 40,000 Hz will bounce back to the module.

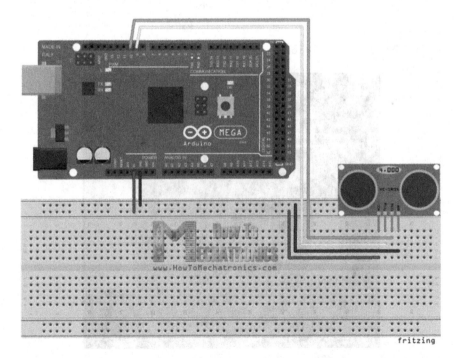

Figure 4.22 Ultrasonic sensor.

The HC-SR04 Ultrasonic Module has 4 pins: Ground, VCC, Trig and Echo. The Ground and the VCC pins of the module needs to be connected to the Ground and the 5 volt pins on the Arduino Board respectively and the trig and echo pins to any Digital I/O pin on the Arduino Board.

The HC-SR04 ultrasonic sensor uses SONAR to determine the distance of an object as shown in Figure 4.23. It offers excellent non-contact range detection with high accuracy and stable readings in an easy-to-use package from 2 cm to 400 cm or 1″ to 13 feet.

For example, if the object is 10 cm away from the sensor, and the speed of the sound is 340 m/s or 0.034 cm/μs, the sound wave will need to travel about 294 u seconds as shown in Figure 4.24. But what you will get from the Echo pin will be double that number because the sound wave needs to travel there and back.

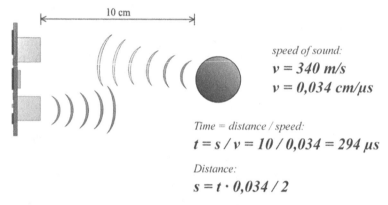

speed of sound:
$$v = 340 \; m/s$$
$$v = 0{,}034 \; cm/\mu s$$

Time = distance / speed:
$$t = s / v = 10 / 0{,}034 = 294 \; \mu s$$

Distance:
$$s = t \cdot 0{,}034 / 2$$

Figure 4.23 Ultrasonic sensor.

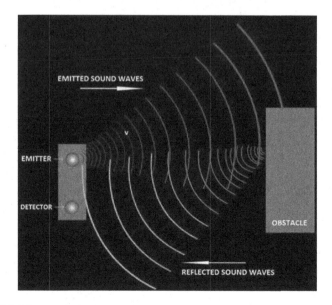

Figure 4.24 Ultrasonic sensor.

Ultrasonic sensors are used primarily as **proximity sensors**. They can be found in automobile self-parking technology and anti-collision safety systems. Ultrasonic sensors are also used in robotic obstacle detection systems, as well as manufacturing technology.

Ultrasonic sensors are also used as **level sensors** to detect, monitor, and regulate liquid levels in closed containers.

- Used in parking sensors, liquid level detection sensors, as well as waste bin sensors.
- Level measurement is a popular application of ultrasonic sensors in detecting liquid and granular materials.
- IOT smart jar: This checks the level of a jar using an ultrasonic sensor and send an alert email to the user. The jar includes an ultrasonic sensor at the top of it and uses the ultra-sonic reflected waves to figure out at what extent the jar is filled and how much space is left inside the jar. Whenever the amount of content changes in the jar, it is sensed by the Node MCU, and the same is updated on the web server. This can be helpful to track supplies and plan for restocking from anywhere in the world.

4.6.12 Acceleration sensors

Acceleration sensors, commonly known as accelerometers, measure the rate at which an object's velocity changes as shown in Figure 4.25. This is in contrast to motion sensors, which measure an object's movement. This modification

Figure 4.25 Accelerometer sensor.

could result from rotational motion, a rapid vibration that causes movement with speed variations, or a free-fall scenario (a directional change). In order to pinpoint an object's location in space and track its movement, an accelerometer monitors the acceleration forces that are acting on the object.

These sensors are used in smartphones, vehicles, aircraft, and other applications to detect the orientation of an object, shake, tap, tilt, motion, positioning, shock or vibration.

One of several technologies that are employed in acceleration sensors include:

- Hall-effect sensors (which rely on measuring changes in magnetic fields).
- Capacitive sensors (which depend on measuring changes in voltage from two surfaces).
- Piezoelectric sensors (which generate a voltage that changes based on pressure from distortion of the sensor).

4.6.12.1 Types of accelerometer

Piezoelectric: used in very high temperatures and high frequency range up to 100 kilohertz.

Piezo-resistive: used in sudden and extreme vibrating applications.

Capacitive: in applications such as a silicon-micro machined sensor material and can operate in frequencies up to 1 kilohertz.

At present, **Micro Electro-Mechanical System (MEMS) Accelerometer** is being used as it is simple, reliable, and highly cost-effective.

Applications

1. In machine monitoring.
2. To measure earthquake activity and aftershocks.
3. Inertial Navigation System (INS): For measuring the position, orientation, and velocity of an object in motion without the use of any external reference. Example: airplane and ship autopilots.
4. In airbag shooting in cars and vehicle stability control.
5. In video games consoles such as the PlayStation 5, to make the steering more controlled, natural and real.
6. In camcorder to stabilize images.
7. Mounted in spacecraft,
8. Mobile Phone: Accelerometers in mobile phones are used to detect the orientation of the phone. The gyroscope, adds an additional dimension to the information supplied by the accelerometer by tracking rotation or twist.
9. Drones: Accelerometers and gyroscopes are the sensors of choice for acquiring acceleration and rotational information in drones.

10. Washing Machines have accelerometers that can detect when the load is out of balance and switch off the electric motor to stop them from spinning themselves to pieces?
11. Electronic Irons and Fan Heaters, have accelerometers inside that detect when they fall over and switch them off to stop them causing fires?

4.6.13 Gyroscopic sensors

Using a three-axis system, gyroscopes (or gyroscopic sensors) are used to measure an object's rotation and calculate its rate of movement, or angular velocity. The orientation of the object can be determined using these sensors without needing to physically view it.

Gyro sensors, also known as angular rate sensors or angular velocity sensors, are devices that sense angular velocity as shown in Figure 4.26.

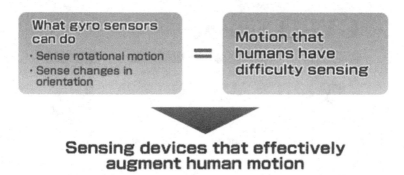

Figure 4.26 Gyroscopic sensor.

It records the rotation/twist about an axis as shown in Figure 4.27 and Figure 4.28.

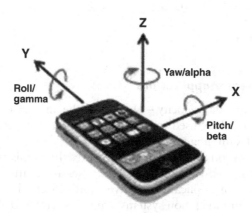

Figure 4.27 Gyroscopic sensor.

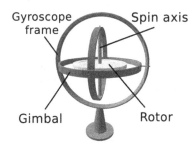

Figure 4.28 Gyroscopic sensor.

4.6.13.1 Types of gyroscope (shown below in Figure 4.29)

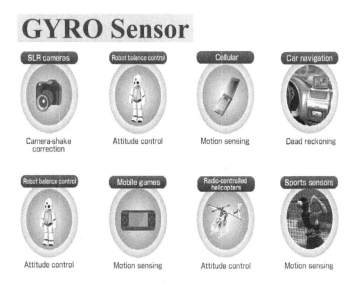

Figure 4.29 Gyro sensor.

4.6.13.2 Gyro sensor applications

1. Sensing of angular velocity – Determine the amount of angular velocity generated. It is used to calculate the actual motion's magnitude. Example: Examining athletic movement.
2. Angle sensing, on the other hand, measures the angular velocity caused by the sensor's own motion. A CPU uses integration operations to find angles. An application receives and reflects the changed angle. Example: Car navigation systems, game controllers, cellular phones.

3. **Control mechanisms–** Senses vibration produced by external factors, and transmits vibration data as electrical signals to a CPU. Used in correcting the orientation or balance of an object. Example: Camera-shake correction, vehicle control.

4.6.13.3 Difference in accelerometer and gyroscope

Accelerometer: It measures **linear acceleration** based on vibration.

Gyroscope: It determines an **angular position** based on the principle of the rigidity of space. Table 4.2 shows the difference between accelerometers and gyroscope.

Table 4.2 Accelerometers vs gyroscope

	Accelerometers	Gyroscopes
What it is	Electromechanical devices that measure acceleration	A device used for measuring rotational changes or maintaining orientation
	Cannot distinguish rotation from acceleration	Unaffected by acceleration
Usage purpose	Measure linear acceleration based on vibration	Measure rate of rotation and angular position around a particular axis
Applications	Commonly found and more applicable in consumer electronics	Commonly found and more applicable in aircrafts, aerial vehicles

4.6.14 Humidity sensors

In order to determine how much water vapor is present in the air or another gas, humidity sensors can measure the relative humidity of those gases. In order to produce materials, it is essential to control environmental factors. Humidity sensors allow readings to be obtained and adjustments to be made to minimize rising or falling levels. HVAC systems frequently use this technology to keep targeted comfort levels.

4.6.15 Optical sensors

Optical sensors provide an electrical signal in response to light that is reflected off of an object, which can then be used to detect or measure a state. These sensors detect the interruption or reflection of a light beam brought on by the presence of an object.

The types of optical sensors include:

- Through-beam sensors (which detect objects by the interruption of a light beam as the object crosses the path between a transmitter and remote receiver)

- Retro-reflective sensors (which combine transmitter and receiver into a single unit and use a separate reflective surface to bounce the light back to the device)
- Diffuse reflection sensors (which operate similarly to retro-reflective sensors except that the object being detected serves as the reflective surface)

4.7 ACTUATORS

An actuator is a device that actuates or moves something as shown in Figure 4.30. An actuator uses some type of energy to provide motion or to apply a force. For example, an electric motor uses electrical energy to create a rotational movement or to turn on object, or to move an object. A tire jack or screw jack uses mechanical energy to provide enough force to lift a car. In short, an actuator converts some type of energy into motion. Actuators include motors, gears, pumps, pistons, valves, and switches

- The actuator is a device that transforms a certain form of energy into motion.
- An actuator basically requires some kind of a control signal and a source of energy.

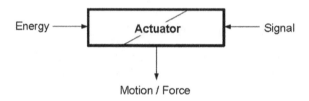

Figure 4.30 Actuators.

Basic Concepts of Actuators: An actuator is something that actuates or moves something. More specifically, an actuator is a device that coverts an input energy into motion or mechanical energy. The input energy of actuators can be "manual" (e.g., levers and jacks), hydraulic or pneumatic (e.g., pistons and valves), thermal (e.g., bimetallic switches or levers), and electric (e.g., motors and resonators). In the transducers unit, a transducer was defined as any device that converts one form of energy to another form of energy; therefore, by that definition, an actuator can be a specific type of a transducer. The motor is one such actuator. A motor converts electrical energy to mechanical energy; therefore, a motor is both an actuator and a transducer.

4.7.1 Hydraulic actuators

Hydraulic actuators have a cylinder or fluid motor that uses hydraulic power to generate mechanical motion, which in turn leads to linear, rotatory, or oscillatory motion. Given the fact that liquids are nearly impossible to compress, a hydraulic actuator can exert a large force. When the fluid enters the lower chamber of the actuator's hydraulic cylinder, the pressure inside increases and exerts a force on the bottom of the piston, also inside the cylinder. The pressure causes the sliding piston to move in a direction opposite to the force caused by the spring in the upper chamber, making the piston move upward and opening the valve. The downside with these actuators is the need for many complementary parts and possibility of fluid leakage An example of hydraulic actuators is JCB machine shown in Figure 4.31.

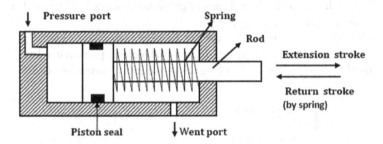

Figure 4.31 JCB machine.

4.7.1.1 Applications of hydraulic systems: five categories

Industrial: Plastic-processing machineries, steel-making and primary metal extraction applications, automated production lines, machine tool industries, paper industries, loaders, crushes, textile machinery, R & D equipment, and robotic systems etc.

Mobile hydraulics: Tractors, irrigation system, earthmoving equipment, material handling equipment, commercial vehicles, tunnel boring equipment, rail equipment, building and construction machineries and drilling rigs etc.

Automobiles: It is used in the systems such as brakes, shock absorbers, steering system, wind shield, lift and cleaning etc.

Marine applications: It mostly covers ocean-going vessels, fishing boats, and navel equipment.

Aerospace equipment: There are equipment and systems used for rudder control, landing gear, brakes, flight control, and transmission etc. which are used in airplanes, rockets, and spaceships.

4.7.2 Pneumatic actuators

Pneumatic actuators convert energy, in the form of compressed air, into mechanical motion as shown in Figure 4.32. Here pressurized gas or compressed air enters a chamber, thereby building up the pressure inside. Once this pressure goes above the required pressure levels in contrast to the atmospheric pressure outside the chamber, it makes the piston or gear move kinetically in a controlled manner, thus leading to a straight or circular mechanical motion. Examples include pneumatic cylinders, air cylinders, and air actuators. Cheaper and often more powerful than other actuators, they can quickly start or stop as no power source has to be stored in reserve for operation. Often used with valves to control the flow of air through the valve, these actuators generate considerable force through relatively small pressure changes.

Examples of maker projects using pneumatic actuators include lifting devices and humanoid robots with arms and limbs, typically used for lifting.

- Example 1: The rack and pinion pneumatic actuators are used for the valve controls on water pipes.
- Example 2: Pneumatic brakes can be very responsive to small changes in pressure that are applied by the driver. So, pneumatic brakes are used in trucks. The hydraulic brakes are more common in cars.

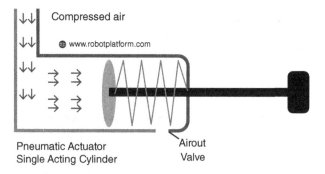

Figure 4.32 Pneumatic actuators.

- Pneumatic and Hydraulic Dangers
- Pneumatic Dangers
- Air Embolism
- Hose/Pipe Whipping
- Noise
- Crushing/Cutting
- Hydraulic Dangers
- High-Pressure Oil Injection
- Oil Burns
- Crushing/Cutting

4.7.2.1 Electric linear actuators

Now let us consider the electrical actuators, those that run on electricity. Taking off from the two basic motions of linear and rotary, actuators can be classified into these two categories: linear and rotaryas shown in Figure 4.33. Electric linear actuators take electrical energy and turn it into straight line motions, usually for positioning applications, and they have a push-and-pull function. They convert energy from the power source into linear motion using mechanical transmission, electromagnetism, or thermal expansion; they are typically used whenever tilting, lifting, pulling, and pushing are needed. They are also known for offering precision and smooth motion control; this is why they are used in industrial machinery, in computer peripherals such as disk drives and printers, opening and closing dampers, locking doors and for braking machine motions. They are also used in 3D printers and for controlling valves. Some of them are unpowered and manually operated with a rotating knob or handwheel.

Figure 4.33 AC motor, stepper motor, DC motor.

4.7.3 Thermal actuators

A thermal actuator is a non-electric motor that generates linear motion in response to temperature changes as shown in Figure 4.34. Its main components are a piston and a thermal-sensitive material. When there is a rise in temperature, the thermal-sensitive materials begin to expand in response, driving the piston out of the actuator. Similarly, upon detecting a drop in the temperature, the thermal-sensitive materials inside contract, making the piston retract. Thus these actuators can be used for carrying out tasks such as releasing latches, working switches and opening or closing valves. They have many applications, particularly in the aerospace, automotive, agricultural and solar industries.

4.7.3.1 Thermal actuators operating principle

A thermostatic actuator consists of a temperature-sensing material. It is this material that expands and contracts based on the temperature of the device, causing the piston to move.

"Cold Position" - Solid State

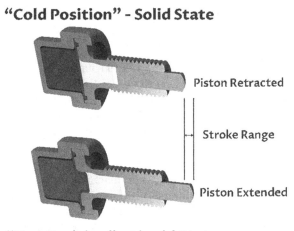

Piston Retracted

Stroke Range

Piston Extended

"Hot Position" - Liquid State

Figure 4.34 Thermal actuator.

4.7.4 Electromechanical actuators

Electromechanical actuators are mechanical actuators where there's an electric motor in place of the control knob or handle as shown in Figure 4.35. The rotary motion of the motor leads to linear displacement. The inclined plane concept is what drives most electromechanical actuators; the lead screw's threads work like a ramp converting the small rotational force by magnifying it over a long distance, thus allowing a big load to be moved over a small distance. While there are many design variations among the electromechanical actuators available today, most have the lead screw and the nut incorporated into the motion. The biggest advantages are their greater accuracy in relation to pneumatics, their longer lifecycle and low maintenance effort required. On the other hand, they do not boast the highest speed.

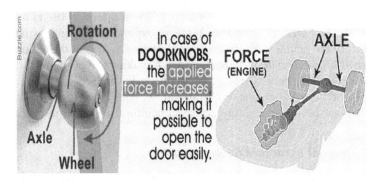

Figure 4.35 Electromechanical actuators.

Types of Motors:

1. Hydraulic motors
2. Pneumatic motors
3. Clutch/brake motor
4. Stepper motors (DC motor)
5. AC motors
6. Servomotors (DC motor)

4.7.4.1 Hydraulic motors

Hydraulic motors move a piston through a tube using pressurized fluid. Hydraulic motors output linear, rotary, or oscillating motion, but acceleration is limited. Hydraulic motors are typically inefficient, can be a fire hazard and require more than usual maintenance.

4.7.4.2 Pneumatic motors

- Pneumatic motors are air-driven, using either vacuum or compressed air, which converts energy into linear or rotary motion as shown in Figure 4.36.
- Air pressure and flow determine speed.

Figure 4.36 Pneumatic motors.

The Illinois-based tech company Bimba adopted the Intellisense platform to drive its pneumatic components as shown in Figure 4.37.

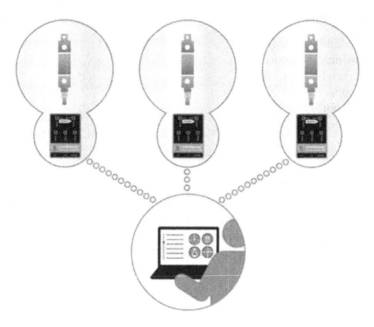

Figure 4.37 Pneumatic motors.

- Three actuators are monitored remotely in the Bimba Intellisense platform.
- Intellisense incorporates sensors, cylinders, and software that enhance productivity through the use of real-time data, providing feedback on pneumatic wear and tear that can be utilized to maximize efficiency.
- This is a prime example of how IoT is driving Industry 4.0 "intelligent" production systems that can communicate in real time with detailed and predictive information on any particular part.

4.7.4.3 Clutch/brake motor

- A clutch/brake motor functions by coupling a continuously rotating motor shaft with a load, stopping only when the load is uncoupled.
- While this motor is easy to apply, relatively inexpensive, and great for light loads, its acceleration is uncontrolled as well as inaccurate.

4.7.4.4 Stepper motors (DC motor)

- A stepper motor is a brushless electromechanical device that converts electrical energy into mechanical energy. It divides rotational motion into an equal number of steps with the help of actuators, depending on the application requirement as shown in Figure 4.38. There are many types of stepper motors such as linear stepper motor, and are made based on the number of steps per revolution. For motion, they make use of actuators. Stepper motors offer consistent repetition of movement, speed control, and precision in positioning. They have a number of industrial applications such as 3D printing, robotics, CNC machining, medical imaging, and so on.

- Stepper motors (DC motor) are electromechanical, converting a digital pulse into rotational movement or displacement.
- While stepper motors are not good for varying loads and are typically not energy-efficient, they are great for constant loads and positional accuracy.
- Examples: vacuum cleaner, hairdryer, elevators, electric windows in cars, etc.

Figure 4.38 130 DC motor.

4.7.4.5 AC motor

A motor that uses AC power is called an AC motor as shown in Figure 4.39.

Figure 4.39 AC motor.

4.7.4.6 Servo motors

This is a rotary actuator that allows for the precise control of angular position. This makes them suitable for use in closed-loop systems where precise position control is needed.

Servo motors are part of a closed-loop system and they are a self-contained electrical device that rotates parts of a machine with high efficiency with great precision as shown in Figure 4.40.

Servo motors are made up of:

• Control circuit
• Small DC motor
• Potentiometer

Figure 4.40 Servo motors.

Servo motors should be used in fields that require high precision such as:

• Machine tools
• Packaging equipment
• Textile equipment
• Laser processing equipment
• Robots
• Automated production lines etc.

Chapter 5

Electronic components used in IoT

5.1 ELECTRONICS COMPONENTS

The Internet of Things, or IoT, refers to any electronic device that is remotely connected to the internet. IoT technology has been integrated into every aspect of our modern world. Electronic systems are used in everything from computers to industrial manufacturing plants, agricultural sciences and even running the most basic devices in the home, such as lighting. Many of these technologies rely on their ability to communicate with their operators remotely through desktop computers or other mobile devices, like smartphones and tablets.

As the electronic technology field expands rapidly, more businesses than ever before are going after a piece of the market share. High-quality, reliable and robust electronic components in the design of all IoT devices ensure the longevity of the product and overall customer satisfaction.

IoT devices designed with high-quality electronic components have long-term, reliable performance that will ultimately steal the market share. Quality is key and consumers make purchasing decisions largely based on product reviews.

5.1.1 Breadboard

A breadboard is a solder-less construction base used for developing an electronic circuit and wiring for projects with microcontroller boards like Arduino. A breadboard is categorized as a **solder-less board**. This means that the component does not require any soldering to fit into the board. Thus, we can say that breadboard can be reused. We can easily fit the components by plugging their end terminal into the board. Hence, a breadboard is often called a **plug-board**. As common as it seems, it may be daunting when first getting started with using one. The term "breadboard" derives from a literal piece of wood used to cut bread, on which, back in the early days, people would build electronic circuits. However, over the years, the design has changed. Now, thanks to the invention by Ronald J. Portugal, the breadboard we know comes in a smaller, more portable white plastic and pluggable design.

The breadboard is a white rectangular board with small, embedded holes in which electronic components can be insertedas shown in Figure 5.1. It is commonly used in electronics projects. We can also say that breadboard is a prototype that acts as a construction base for electronics.

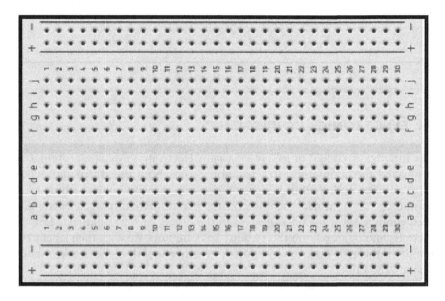

Figure 5.1 Breadboard.

There are three parts in a breadboard, as shown below in Figure 5.2:

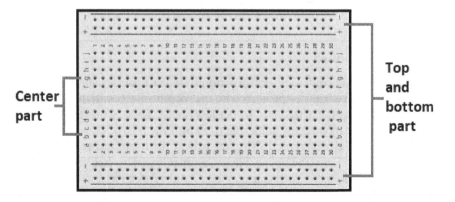

Figure 5.2 Parts of breadboard.

The top and bottom holes of a row in a breadboard are connected horizontally, and the center part is connected vertically, as shown in Figure 5.3:

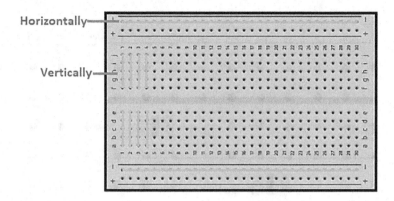

Figure 5.3 Breadboard.

It means a single horizontal line of a breadboard has the same connection. It is because the metal strip underneath the breadboard at the top and bottom are connected horizontally. Hence, it provides the same connection in a row. The two top and bottom parts of a breadboard are generally used for power connections.

The vertical connection of the center part means a single vertical line in a breadboard provides the same connection. It is useful when we need to connect the different components in series in Figure 5.3.

For example, let's connect two resistors in series. These two resistors can be connected in series in different ways, as shown below in Figure 5.4:

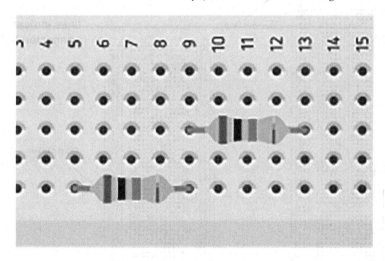

Figure 5.4 Breadboard.

This is because the metal strips underneath the breadboard at the center are connected vertically. Hence, it provides similar connectivity through a particular column, as shown in Figure 5.5:

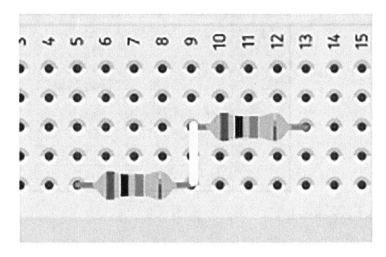

Figure 5.5 Breadboard.

The connection between two different components can be created by inserting a lead in common in Figure 5.6. For example, a jump wire which acts as a connection between the LED and battery terminal can be connected in any hole in the same vertical line.

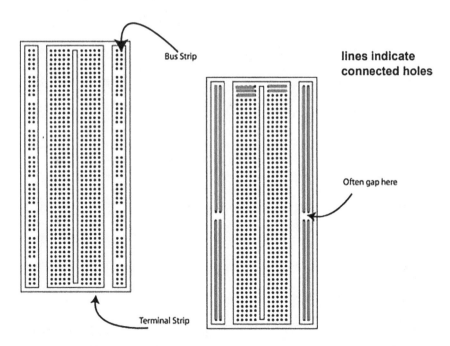

Figure 5.6 Breadboard.

Bus strips are mainly used for power supply connections. Terminal strips are principally used for electrical components. The holes colored in orange are connected together. These sets of connecting holes can be called a node, where it's possible to interconnect the node from bus strips to terminal strips with jumper wires.

5.1.1.1 How to read breadboard rows and columns?

These are written to help you locate the individual hole in the breadboard, in a manner similar to how finding a cell in an Excel spreadsheet works. The example as seen below: Hole C12 = Column C, Row 12.

The positive and negative signs on both sides of the breadboard are power rails, used to power your circuit by connecting battery pack or external power supply as shown in Figure 5.7.

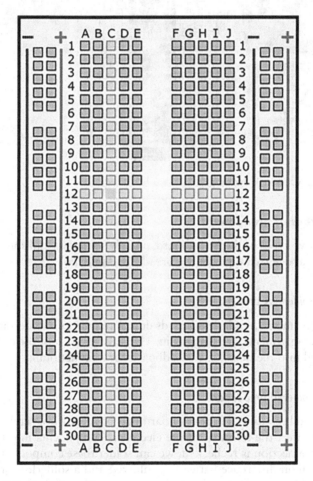

Figure 5.7 Breadboard reading.

Columns at the edges are connected from top to bottom inside the breadboard generally used for supply and ground. Inside the breadboard, the holes in each row are connected up to the break in the middle of the board as shown in Figure 5.8.

For example, A1, B1, C1, D1, and E1 all have a wire inside of the breadboard to connect them. Then F1, G1, H1, I1, and J1 are all connected, but A1 is not connected to F1.

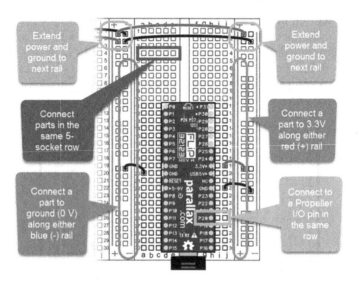

Figure 5.8 Breadboard reading.

5.1.1.2 Types of breadboard

There are two types of the breadboard, namely solderless and soldered. Let's discuss the above two types of the breadboard in detail.

5.1.1.2.1 Solderless breadboards

As the name implies, solderless boards do not require any soldering after the electronic components are plugged in. The leads or ends of the components are inserted into the holes of a breadboard for its functioning.

5.1.1.2.2 Soldered breadboard

The soldered breadboard is also a board that has a tiny hole embedded into it. We can insert the terminal of the electronic components into the board. After the connection is rechecked, we can solder these components.

The common difference between a soldered and a solderless breadboard is the **reusability**.

5.1.1.2.3 Connection setup through a breadboard

Here, we will discuss the connection setup through the breadboard with the help of one example. The will help us to understand the different connections involved.

Example: Blinking an LED

The components required for the above example are an LED (any color), a breadboard, two jump wires, a resistor, and a battery. Here, we have chosen a red LED. The resistor is connected in series with the LED to limit the current across the LED. LED has two terminals, namely cathode (negative terminal) and anode (positive terminal). The structure of LED is shown below in Figure 5.9:

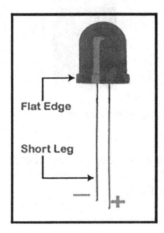

Figure 5.9 LED blinking.

Connection Setup: The connection setup is listed in the below steps and also shown in Figure 5.10:

Plug in the two terminals of the LED into the two tiny holes of the breadboard. We can add or skip the resistor because the battery already provides a limiting current. Connect one end of a jump wire to the anode of an LED and the other end of the jump wire to the positive terminal of the battery. Similarly, connect the cathode of an LED to the negative terminal of the battery. As soon as the circuit is complete after the terminals are connected to the battery, the LED will light. The circuit thus formed will appear like the image shown here:

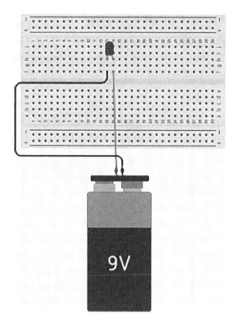

Figure 5.10 Connection setup in breadboard.

The above circuit connection depicts that the LED and resistor are connected in series. Similarly, we can easily create various projects and circuits with the help of the breadboard.

Advantages of a Breadboard:

The advantages of using a breadboard are as follows:

- Temporary prototype: We can build a temporary prototype for the projects with the help of a breadboard.
- Reusable: Today, solderless boards are mostly used in various applications. This does not require any soldering to fix the components. Hence, it can be reused.
- Lightweight: The breadboard is made of white plastic, which is light in weight.
- Easy experimentation: We can quickly insert the leads of the components into the tiny holes of the breadboard. The circuit can be created using various components and circuit design.
- Inexpensive: The breadboards are easily available. It also cost less.
- Easy to use: It does not involve any complex parts. We can easily insert the required number of components.

- No drilling required: The holes are already embedded in the board. Hence, we do not require any drilling to insert the electronic components.
- Quick modifying capability: We can easily switch or remove the components from the board.
- Available in various sizes: The breadboards are available in various sizes. We can select the desired size as per the number of components.
- Easy to adjust: The breadboard is easy to adjust in the project or connection setup.

Disadvantages of Breadboard:

- Not suitable for high-current applications.
- Well suited to low-frequency projects.
- Solderless boards are limited to low-frequency applications.
- Requires more physical space for simple circuits.
- A high number of connections in the solderless board make the circuit messy due to the large number of wires.
- The circuit design does not work well for high-speed design.
- The plugging and unplugging can disturb the other connections.
- Less reliable connections.
- Limited signaling.

Alternatives: There are other alternatives to create a prototype for the projects and circuit design. Modern computer systems contain various transistors, resistors, and other electronic components to create a prototype. It does not require any breadboard to build a prototype.

5.1.2 Resistor

A resistor is a passive electrical component with the primary function to limit the flow of the electric current in a circuit, thereby minimizing the huge loss to the electronic devices as shown in Figure 5.11. Resistors resist the flow of electricity and the higher the value of the resistor, the more it resists, and the less electrical current will flow through it. For example, to control how much electricity flows through the LED and therefore how brightly it shines.

Resistors are measured in hundreds of Ohms, thousands of Ohms (kilo Ohms, kΩ), or millions of Ohms (mega Ohms, MΩ). There are different values of resistor, 270 Ω, 470 Ω, 2.2 kΩ, and 10 kΩ. These resistors all look the same, except that they have different-colored stripes on them. These stripes tell you the value of the resistor. It is important to know how to identify the nominal resistance and the tolerance of a resistor. For resistors with +-5% or +-10% tolerance, the color code consists of 4 color bands. For resistors with +-1% or +-2% tolerance, the color code consists of 5 bands.

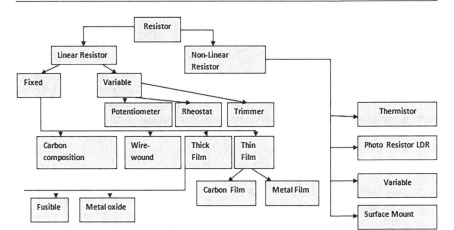

Figure 5.11 Resistor.

Resistance is measured based on the color bands on the resistors. Generally, 4-band, 5-band and 6-band resistors are available. You can calculate the resistance of these resistors using this resistor color code calculator.

The resistor color code works like this and also shown in Table 5.1. For resistors like this with three colored stripes and then a gold stripe at one end. Each color has a number, as follows:

Black	0
Brown	1
Red	2
Orange	3
Yellow	4
Green	5
Blue	6
Purple	7
Gray	8
White	9

Table 5.1 The resistor color code table

Color	Digit	Multiplier	Tolerance
Black	0	1	
Brown	1	10	±1%
Red	2	100	±2%
Orange	3	1,000	
Yellow	4	10,000	

(Continued)

Table 5.1 (Continued) The resistor color code table

Color	Digit	Multiplier	Tolerance
Green	5	100,000	±0.5%
Blue	6	1,000,000	±0.25%
Violet	7	10,000,000	±0.1%
Grey	8		±0.05%
White	9		
Gold		0.1	±5%
Silver		0.01	±10%
None			±20%

Then we can summarize the different weighted positions of each colored band which makes up the resistors color code from Table 5.1 in Table 5.2:

Table 5.2 Colored bands

Number of Colored Bands	3 Colored Bands (E6 Series)	4 Colored Bands (E12 Series)	5 Colored Bands (E48 Series)	6 Colored Bands (E96 Series)
1st Band	1st Digit	1st Digit	1st Digit	1st Digit
2nd Band	2nd Digit	2nd Digit	2nd Digit	2nd Digit
3rd Band	Multiplier	Multiplier	3rd Digit	3rd Digit
4th Band	—	Tolerance	Multiplier	Multiplier
5th Band	—	—	Tolerance	Tolerance
6th Band	—	—	—	Temperature Coefficient

5.1.2.1 Calculating resistor values

The **Resistor Color Code** system is all well and good but we need to understand how to apply it in order to get the correct value of the resistor. The "left-hand", or the most significant colored band, is the band which is nearest to a connecting lead with the color-coded bands being read from left to right as follows:

$$\text{Digit, Digit, Multiplier} = \text{Color, Color} \times 10^{color} \text{ in Ohms} (\Omega)$$

For example, a resistor has the following colored markings:

$$\text{Yellow Violet Red} = 472 = 47 \times 10^2 = 4700 \, \Omega \text{ or 4k7 Ohm.}$$

The fourth and fifth bands are used to determine the percentage tolerance of the resistor. Resistor tolerance is a measure of the resistors' variation from the specified resistive value and is a consequence of the manufacturing process and is expressed as a percentage of its "nominal" or preferred value.

Typical resistor tolerances for film resistors range from 1% to 10%, while carbon resistors have tolerances up to 20%. Resistors with tolerances lower than 2% are called precision resistors with the lower-tolerance resistors being more expensive.

Most five band resistors are precision resistors with tolerances of either 1% or 2% while most of the four band resistors have tolerances of 5%, 10%, and 20%. The color code used to denote the tolerance rating of a resistor is given as:

Brown = 1%, Red = 2%, Gold = 5%, Silver = 10%

If a resistor has no fourth tolerance band then the default tolerance would be at 20%.

It is sometimes easier to remember the resistor color code by using short, easily remembered sentences in the form of expressions, rhymes, and phrases, called an *acrostic*, which have a separate word in the sentence to represent each of the Ten + Two colors.

5.1.2.2 Resistance computation

Minimum Resistance Value: Multiply the nominal value with the tolerance and then subtract this from the nominal value.

Maximum Resistance Value: Multiply the nominal value with the tolerance and then add this from the nominal value.

The first two striped are the first two digits of the value, so red, purple means 2, 7.

The next stripe is the number of zeros that need to come after the first two digits, so if the third stripe is brown, as it is in the photograph above, then there will be one zero and so the resistor is 270 Ω.

Figure 5.12 Resistor.

A simple example is a resistor in series with an LED as shown in Figure 5.12.

Aim – To have a current limiting resistor in series with your LED so that you can control the amount of current through the LED.

If too much current is going through your LED, it will burn out too fast. If too little current is going through it, it might not be enough to light the LED.

E.g. – LED needs 15 mA and has a voltage drop of 2 volts. You have a 5 V power source that you would like to power it with. Which resistor value do you need?

Value for the current limiting resistor is 200 Ohms by using Ohm's Law.

5.1.3 Potentiometer

A potentiometer is a three-terminal resistor where resistance can be manually adjusted to control the current flow. Potentiometer (also known as a pot or potmeter) is defined as a three-terminal variable resistor in which the resistance is manually varied to control the flow of electric current. as shown in Figure 5.13. A potentiometer acts as an adjustable voltage divider. It is a small electronic component whose resistance can be adjusted manually.

Example: controls the rate at which an LED blinks.

We connect three wires to the Arduino board. For example: to control the audio volume of the radio, fan regulator, sliders in a DJ system.

A potentiometer has three pins. Two terminals (the blue and green) are connected to a resistive element and the third terminal (the black one) is connected to an adjustable wiper. The wiper can be moved along the resistive track either by use of a linear sliding control or a rotary "wiper" contact.

Potentiometers can be used as voltage dividers. To use the potentiometer as a voltage divider, all the three pins are connected. One of the outer pins is connected to the GND, the other to Vcc and the middle pin is the voltage output.

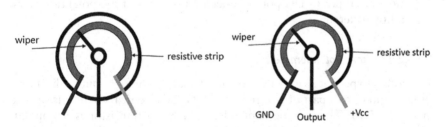

Figure 5.13 Potentiometer.

5.1.3.1 How Does a Potentiometer Work?

A potentiometer is a passive electronic component. Potentiometers work by varying the position of a sliding contact across a uniform resistance. In a potentiometer, the entire input voltage is applied across the whole length of the resistor, and the output voltage is the voltage drop between the fixed and sliding contact, as shown below.

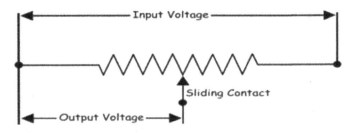

Figure 5.14 Potentiometer.

A potentiometer has the two terminals of the input source fixed to the end of the resistor. To adjust the output voltage, the sliding contact gets moved along the resistor on the output side as shown in Figure 5.14.

5.1.3.2 Potentiometer types

There are two main types of potentiometers:

Rotary potentiometer
Linear potentiometer

Although the basic constructional features of these potentiometers vary, the working principle of both these types of potentiometers is the same. Note that these are types of DC potentiometers: the types of AC potentiometers are slightly different.

5.1.3.2.1 Rotary potentiometers

The rotary-type potentiometers are used mainly for obtaining adjustable supply voltage to a part of electronic circuits and electrical circuits as shown in Figure 5.15. The volume controller of a radio transistor is a popular example of a rotary potentiometer where the rotary knob of the potentiometer controls the supply to the amplifier.

This type of potentiometer has two terminal contacts between which a uniform resistance is placed in a semi-circular pattern. The device also has a middle terminal, which is connected to the resistance through a sliding contact attached with a rotary knob. By rotating the knob one can move the sliding contact on the semi-circular resistance. The voltage is taken between a resistance end contact and the sliding contact. The potentiometer is also named as the POT in short. POT is also used in substation battery chargers to adjust the charging voltage of a battery. There are many more uses of rotary type potentiometer where smooth voltage control is required.

Figure 5.15 Rotary potentiometers.

5.1.3.2.2 Linear potentiometers

The linear potentiometer is basically the same; the only difference is that here instead of rotary movement the sliding contact gets moved on the resistor linearly as shown in Figure 5.16. Here two ends of a straight resistor are connected across the source voltage. A sliding contact can be slide on the resistor through a track attached along with the resistor. The terminal connected to the sliding is connected to one end of the output circuit and one of the terminals of the resistor is connected to the other end of the output circuit.

This type of potentiometer is mainly used to measure the voltage across a branch of a circuit, for measuring the internal resistance of a battery cell, for comparing a battery cell with a standard cell and in our daily life, it is commonly used in the equalizer of music and sound mixing systems.

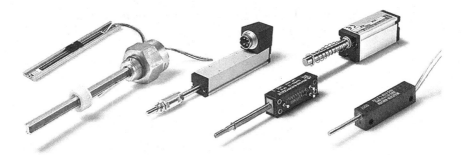

Figure 5.16 Linear potentiometers.

5.1.3.2.3 Digital potentiometers

Digital potentiometers are three-terminal devices, with two fixed end terminals and one wiper terminal which is used to vary the output voltage as shown in Figure 5.17.

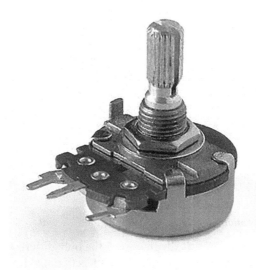

Figure 5.17 Digital potentiometers.

Digital potentiometers have various applications, including calibrating a system, adjusting offset voltage, tuning filters, controlling screen brightness, and controlling sound volume.

However, mechanical potentiometers suffer from some serious disadvantages which make it unsuitable for applications where precision is required. Size, wiper contamination, mechanical wear, resistance drift, sensitivity to vibration, humidity, etc. are among the main disadvantages of a mechanical potentiometer. Hence to overcome these drawbacks, digital potentiometers are more common in applications since it provides higher accuracy.

Motor-Controlled: The speed of the motor and brightness of the LED are controlled by a potentiometer.

Bar Graph: There are ten LEDs connected to an Arduino Uno and a potentiometer. When you rotate the potentiometer, the LEDs will be lit one by one.

Simple LCD Timer With Arduino UNO: The potentiometer is used to choose the contrast of your LCD screen.

5.1.4 PWM

Pulse Width Modulation (PWM) is a technique used in controlling the brightness of LED, the speed control of DC motor, a servo motor or where you must get analog output with digital means.

The Arduino digital pins either gives us 5 V (when turned HIGH) or 0 V (when turned LOW) and the output is a square wave signal. So if we want to dim a LED, we cannot get the voltage between 0 and 5 V from the digital pin, but we can change the ON and OFF time of the signal. If we will change the ON and OFF time fast enough then the brightness of the LED will be changed.

PWM is useful in varying the intensity of a signal such as the brightness of an LED diode, the ping time of sensors, or the power delivery of servomotors as shown in Figure 5.18.

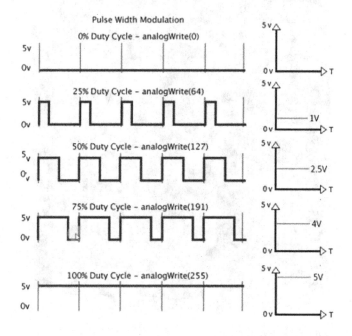

Figure 5.18 PWM.

The Arduino IDE has a built-in function "analogWrite()", which can be used to generate a PWM signal. The frequency of this generated signal for most pins will be about 490 Hz and we can give the value from 0 to 255 using this function.

analogWrite(0) means a signal of 0% duty cycle.
analogWrite(127) means a signal of 50% duty cycle.
analogWrite(255) means a signal of 100% duty cycle.

A 100% duty cycle would mean the output is constantly high, a 50% duty cycle would mean that half the period is high, and half is low, and a 0% duty cycle would mean that the signal is constantly off.

Arduino Uno has 6 8-bit PWM channels. The pins with symbol '~' show that it has PWM support.

The PWM pins are 3, 5, 6, 9, 10 and 11 as shown in Figure 5.19.

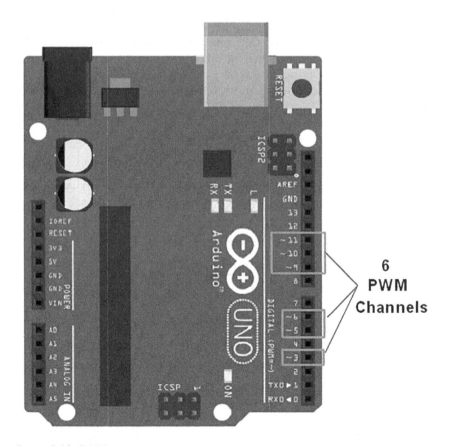

Figure 5.19 PWM pins.

5.1.5 Jumper wire

Jumper wires are extremely handy components to have on hand, especially when prototyping . Simply wires that have connector pins at each end, allowing them to be used to connect two points to each other without soldering.

Used with breadboards and other prototyping tools in order to make it easy to change a circuit as needed.

Basically, we have three types as shown in Figure 5.20:

1. Male to male
2. Female to female
3. Male to female

Figure 5.20 Jumper wires.

5.1.6 Arduino

Arduino acts as the brain of the system and processes the data from the sensor. Arduino is an open-source hardware platform that is readily available for hobbyists & enthusiasts across the globe to build projects. It comes with an ATMega microcontroller that processes the data and facilitates the proper working of the IoT system. And the beauty is that the Arduino can be programmed 'n' number of times, making it possible for you to build various types of IoT projects just by changing a simple code. The Arduino is an open-source electronics platform based on easy-to-use hardware and software used to build electronics projects. All Arduino boards have one thing in common, which is a microcontroller. A microcontroller is basically a really small computer.

With the Arduino, you can design and build devices that can interact with your surroundings. The Arduino boards are basically a tool for controlling electronics. They are able to read inputs with their onboard microcontroller (e.g. light on a sensor, an object near a sensor) and turn it into an output (drive a motor, ring an alarm, turning on an LED, display information on an LCD).

With the Arduino, makers and electricians can easily prototype their products and make their ideas come to life.

5.1.6.1 Why use the Arduino?

There are many electronic boards out there, so why use the Arduino board? Well, there are many reasons that make this microcontroller special. The advantages of using the Arduino include:

- Arduino simplifies microcontrollers for beginners.
- Besides the main microcontroller chip, a microcontroller will require many different parts for it to work. What Arduino did is that they took all the essential components of a microcontroller and design it in a way that is very simple to operate of a piece of PCB. This makes the Arduino boards welcoming to all beginners!
- Furthermore, with Arduino easy-to-use IDE software for beginners, the Arduino are easier to learn to program as it uses a simplified version of C++ compared to other programming software. Because of this, the Arduino is commonly cited as the pathway for everyone who is looking to learn about microcontrollers. With it being optimized for users of all levels, even advanced users are taking advantage of the Arduino IDE as well!
- In addition, the Arduino community is very big and many users and organizations are all using it. Cheap.
- Whenever you are buying something, you will always look at the cost first. The Arduino are very accessible and cost-effective!

Arduino IDE is also cross-platform which means you can run it on Windows, Macintosh OSX, and also Linux operating systems making it much more flexible than when compared with other microcontroller systems, which can only run Windows.

5.1.6.2 Wide variety

The Arduino has many variations for you to choose from to allow you to pick one that suits your project the most!

Having space constraints? You can get yourself an Arduino Nano, which is only 43.18 mm by 18.54 mm! Require more memory space and processing power? You can get yourself an Arduino Mega!

We will talk more about all the different types of Arduino's and their differences later on.

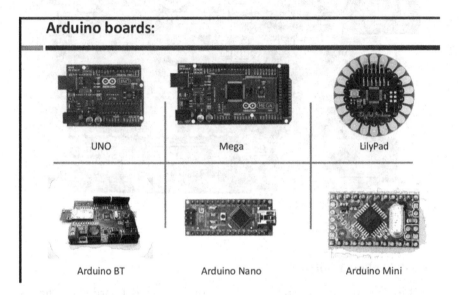

Arduino boards:

UNO   Mega   LilyPad

Arduino BT   Arduino Nano   Arduino Mini

Figure 5.21 Arduino board.

5.1.6.3 Arduino UNO

The development of the Arduino UNO board is considered as new compared to other Arduino boards. This board comes up with numerous features that help the user to use this in their project. The Arduino UNO uses the Atmega16U2 microcontroller, which helps to increase the transfer rate and contain large memory compared to other boards. No extra devices are needed for the Arduino UNO board, such as joystick, mouse, keyboard and many more. The Arduino UNO contains SCL and SDA pins and also have two additional pins fit near to RESET pin as shown in Figure 5.21.

The board contains 14 digital input pins and output pins in which 6 pins are used as PWM, 6 pins as analog inputs, USB connection, reset button and one power jack as shown in Figure 5.22. The Arduino UNO board can be attached to the computer system buy USB port and also get a power supply to board from the computer system. The Arduino UNO contains flash memory of size 32 KB that is used to the data in it. The other feature of the Arduino UNO is compatibility with other shield and can be combined with other Arduino products.

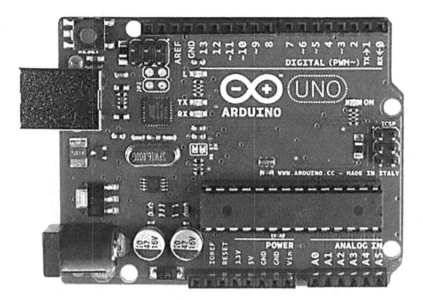

Figure 5.22 Arduino UNO.

5.1.6.4 LilyPad Arduino

The LilyPad Arduino is considered as other Arduino board type that is designed for integrating with wearable projects and e-textile projects. This board comes in a round shape as shown in Figure 5.23 that helps to decrease the snagging and can be easily connected to other devices. This board uses the ATmega328 microcontroller and Arduino bootloader in it. This board uses very less external component in it that makes the design easy and compatible.

The board requires a 2 volt to 5 volt power supply and uses large pin holes so that it can be easily connected to other devices. This board is widely used for controlling different devices which includes a motor, light, and switch. The components of this board, such as the sensor board, the input board, and the output board, can be washable because this board is used in clothing industries.

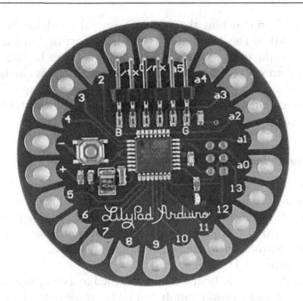

Figure 5.23 Lilypad Arduino.

5.1.6.5 Arduino Mega

This board is considered as the microcontroller that uses the AtMega2560 in it as shown in Figure 5.24. There are total of 54 input pins and output pins in it in which 14 pins are of PWM output, 4 pins are of hardware port, 16 pins as analog inputs. The board also contain one USB connection, ICSP header, power jack and one REST pin.

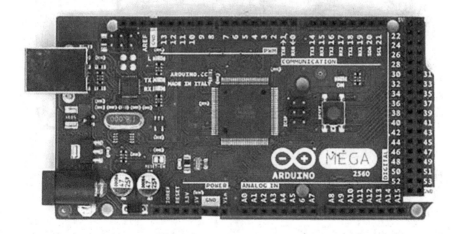

Figure 5.24 Arduino Mega.

There are additional pins that act as a crystal oscillator, having a frequency of 16 MHz. The board also has flash memory of 256 KB size which is used to store the data. The Arduino Mega board can be attached to the computer system via a USB connection and a power supply can be provided to the board by using either a battery or an AC to DC adapter. The board also has a large number of pins fitted into it, making it suitable for more complex projects.

5.1.6.6 Arduino leonardo

This board is considered as the microcontroller that uses the AtMega32u4 in it as shown in Figure 5.25. There are a total of 20 digital input pins and output pins, ofwhich7 pins are used as PWM and 12 pins are used as analog inputs. The board also contains one micro USB connection, a power jack, and a RESET button. There are additional pins which act as crystal oscillators at a frequency of 16 MHz.

The Arduino Leonardo board can be attached to a computer system via a USB connection and a power supply can be provided to the board by using a battery or an AC to DC adapter. The microcontroller used by the Arduino Leonardo has an in-built USB connection that removes any dependency on an extra processor. As there is no additional USB connection in the board, it helps the board to act as a mouse or a keyboard for the computer system. The Arduino Leonardo is considered to be the cheapest of the Arduino boards.

Figure 5.25 Arduino Leonardo.

5.1.6.7 Arduino Red board

The Arduino Red board is another type of Arduino board that uses the mini USB cable for getting programmed and the Arduino IDE is used for this purpose. This board is compatible with the Windows 8 operating system and there is no need to change the security settings to make this board work. The Red board uses the FTDI chip and the USB chip to connect to other devices. As the design of the Red board is very simple, it can easily be integrated with other projects. The only requirement is to plug in the Red board and select the appropriate option. It can then upload a program very quickly. The barrel jack can be used to control the USB cable of the Arduino Red board.

5.1.6.8 Arduino shields

Arduino shields are considered to be pre-built circuit boards that are used to connect other Arduino boards. An Arduino shield is placed on top of an Arduino boards and this enhances the capability of the board to be connected to the Internet network, to control a motor or an LCD, or to establish wireless communication as shown in Figure 5.26. There are different type of shields available to use. These include wireless shields, Ethernet shields, proto Shields, and GSM shields. The availability of all these different types helps to increase the compatibility of Arduino boards.

Figure 5.26 Arduino shields.

5.1.6.9 Arduino Nano

The Arduino Nano is a small Arduino board based on an ATmega328P or an ATmega628 microcontroller. The connectivity is the same as the Arduino UNO board.

Table 5.3 Arduino board names

Board Name	Operating Volt (V)	Clock Speed (MHz)	Digital i/o	Analog Inputs	PWM	UART	Programming Interface
Arduino Uno R3	5	16	14	6	6	1	USB via ATMega16U2
Arduino Uno R3 SMD	5	16	14	6	6	1	USB via ATMega16U2
Red Board	5	16	14	6	6	1	USB via FTDI
Arduino Pro 3.3 v/8 MHz	3.3	8	14	6	6	1	FTDI-Compatible Header
Arduino Pro 5 V/16 MHz	5	16	14	6	6	1	FTDI-Compatible Header
Arduino Mini 05	5	16	14	8	6	1	FTDI-Compatible Header

The Nano board is defined as a sustainable, small, consistent, and flexible microcontroller board. It is small in size compared to the UNO board. The devices required to start our projects using the Arduino Nano board are Arduino IDE and mini USB as shown in Table 5.3.

The Arduino Nano includes an I/O pin set of 14 digital pins and 8 analog pins as shown in Figure 5.27. It also includes 6 power pins and 2 reset pins.

Figure 5.27 Arduino Nano.

5.1.7 Arduino UNO

The Arduino UNO is a standard board of Arduino. Here UNO means 'one' in Italian. It was named UNO to label the first release of Arduino Software. It was also the first USB board released by Arduino. It is considered to be a powerful board which can be used in various projects. Arduino.cc developed the Arduino UNO board. Arduino UNO is based on an ATmega328Pmicrocontroller. It is easy to use compared to other boards, such as the Arduino Mega board, etc. The board consists of digital and analog Input/Output pins (I/O), shields, and other circuits.

The Arduino UNO includes 6 analog pin inputs, 14 digital pins, a USB connector, a power jack, and an ICSP (In-Circuit Serial Programming) header as shown in Figure 5.28. It is programmed based on IDE, which stands for Integrated Development Environment. It can run on both online and offline platforms. The IDE is common to all available boards of Arduino.

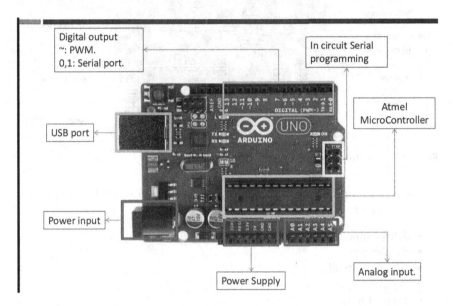

Figure 5.28 Arduino UNO.

Arduino boards can read analog or digital input signals from different sensors and turn it into an output, such as activating a motor, turning a LED on or off, connecting to the cloud and many other actions.

You can control your board functions by sending a set of instructions to the microcontroller on the board via Arduino IDE (referred to as uploading software).

Unlike most previous programmable circuit boards, Arduino does not need an extra piece of hardware (called a programmer) to load a new code onto the board. You can simply use a USB cable.

Additionally, the Arduino IDE uses a simplified version of C++, making it easier to learn to program.

Finally, Arduino provides a standard form factor that breaks the functions of the microcontroller into a more accessible package.

Arduino boards based on ATMEGA328 microcontroller

UNO is not the only board in the Arduino family. There are other boards like Arduino Lilypad, Arduino Mini, Arduino Mega, and Arduino Nano.

However, the Arduino UNO board became more popular than other boards in the family because it has documentation that is much more detailed. This led to its increased adoption for electronic prototyping, creating a vast community of electronic geeks and hobbyists.

In recent times, the UNO board has become synonymous with Arduino.

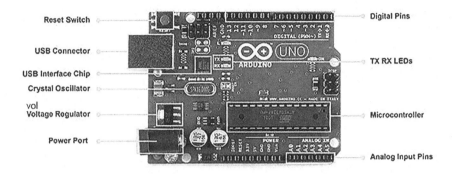

Figure 5.29 Arduino board pins.

The major components of Arduino UNO board are as follows and also shown in Figure 5.29:

- USB connector
- Power port
- Microcontroller
- Analog input pins
- Digital pins
- Reset switch
- Crystal oscillator
- USB interface chip
- TX RX LEDs

Arduino UNO R3 description in Figure 5.30:

Figure 5.30 Arduino board description.

- Microcontroller Microchip ATmega328P
- Operating Voltage 5 V
- USB Standard Type B
- Digital I/O Pins 14
- PWM Digital I/O Pins 6
- Analog Input Pins 6
- Flash Memory 32 KB
- SRAM 2 KB
- EEPROM 1 KBClock Speed 16 MHz
- USB 2.0 Printer Cable is A-Male to B-Male Cord: One 6-foot-long (1.8 meters) high-speed multi-shielded USB 2.0 A-Male to B-Male cable
- Connects mice, keyboards, and speed-critical devices, such as external hard drives, printers, and cameras to your computer
- Constructed with corrosion-resistant, gold-plated connectors for optimal signal clarity and shielding to minimize interference
- Full 2.0 USB capability/480 Mbps transfer speed

Now let's take a closer look at each component.

5.1.7.1 USB Connector

This is a printer USB port used to load a program from the Arduino IDE onto the Arduino board. The board can also be powered through this port as shown in Figure 5.31.

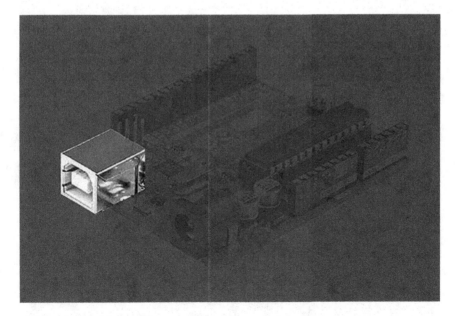

Figure 5.31 USB connector.

5.1.7.2 Power Port

The Arduino board can be powered through an AC-to-DC adapter or a battery as shown in Figure 5.32. The power source can be connected by plugging in a 2.1 mm center-positive plug into the power jack of the board.

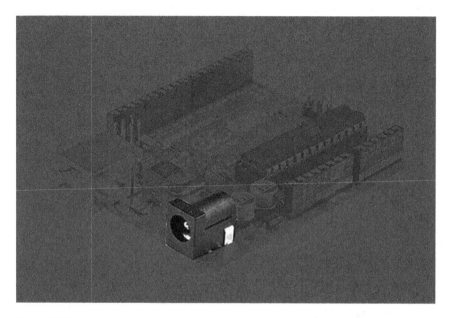

Figure 5.32 Power port.

The Arduino UNO board operates at a voltage of 5 volts, but it can withstand a maximum voltage of 20 volts. If the board is supplied with a higher voltage, there is a voltage regulator (it sits between the power port and USB connector) that protects the board from burning out as shown in Figure 5.33.

Figure 5.33 1 mm center-positive plug.

5.1.7.3 Microcontroller

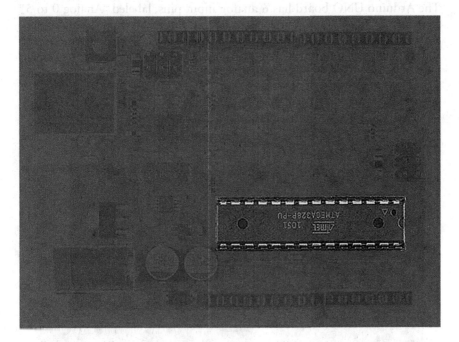

Figure 5.34 Microcontroller.

5.1.7.3.1 Atmega328P microcontroller

It is the most prominent black rectangular chip with 28 pins. Think of it as the brains of your Arduino. The microcontroller used on the UNO board is Atmega328P by Atmel (a major microcontroller manufacturer). Atmega328P contains the following components as shown in Figure 5.34:

- **Flash memory** of 32 KB. The program loaded from Arduino IDE is stored here.
- **RAM** of 2 KB. This is a runtime memory.
- **CPU**: It controls everything that goes on within the device. It fetches the program instructions from flash memory and runs them with the help of RAM.
- **Electrically Erasable Programmable Read Only Memory (EEPROM)** of 1 KB. This is a type of nonvolatile memory, and it keeps the data even after device restart and reset.

Atmega328P is pre-programmed with bootloader. This allows you to directly upload a new Arduino program into the device, without using any external hardware programmer, making the Arduino UNO board easy to use.

5.1.7.4 Analog input pins

The Arduino UNO board has 6 analog input pins, labeled "Analog 0 to 5" as shown in Figure 5.35. These pins can read the signal from an analog sensor such as a temperature sensor and convert it into a digital value so that the system understands. These pins just measure voltage and not the current because they have very high internal resistance. Hence, only a small amount of current flows through these pins.

Although these pins are labeled analog and are analog input by default, they can also be used for digital input or output.

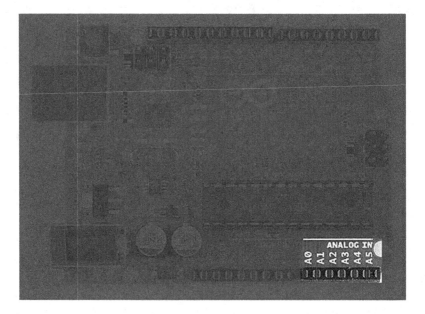

Figure 5.35 Analog input pins.

5.1.7.5 Digital pins

You can find these pins labeled "Digital 0 to 13" as shown in Figure 5.36. These pins can be used as either input or output pins. When used as output, these pins act as a power supply source for the components connected to them. When used as input pins, they read the signals from the component connected to them.

When digital pins are used as output pins, they supply 40 milliamps of current at 5 volts, which is more than enough to light an LED.

Some of the digital pins are labeled with tilde (~) symbol next to the pin numbers (pin numbers 3, 5, 6, 9, 10, and 11). These pins act as normal digital pins but can also be used for Pulse-Width Modulation (PWM), which simulates analog output like fading an LED in and out.

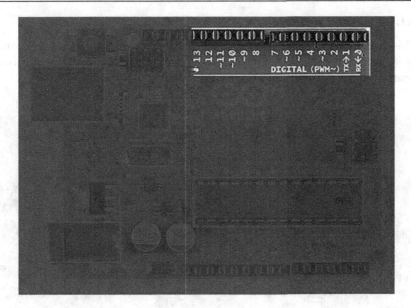

Figure 5.36 Digital pins.

5.1.7.6 Reset switch

When this switch is clicked, it sends a logical pulse to the reset pin of the microcontroller, and now runs the program again from the start as shown in Figure 5.37. This can be very useful if your code doesn't repeat, but you want to test it multiple times.

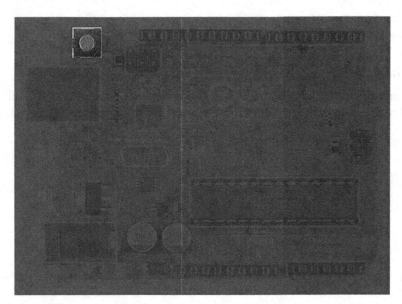

Figure 5.37 Reset switch.

5.1.7.7 Crystal oscillator

This is a quartz crystal oscillator which ticks 16 million times a second as shown in Figure 5.38. On each tick, the microcontroller performs one operation, for example, addition, subtraction, etc.

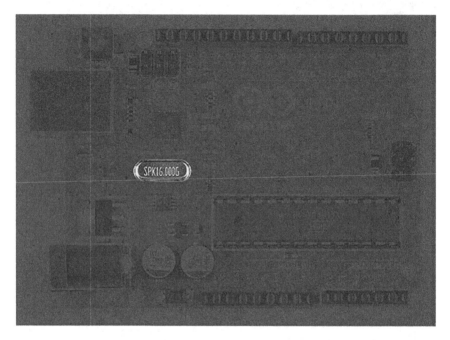

Figure 5.38 Crystal oscillator.

5.1.7.8 USB interface chip

Think of this as a signal translator. It converts signals in the USB level to a level that an Arduino UNO board understands as shown in Figure 5.39.

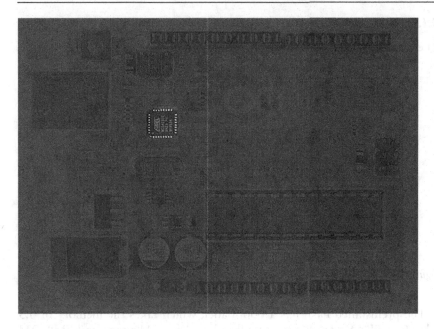

Figure 5.39 USB interface chip.

5.1.7.9 TX–RX LEDs

TX stands for transmit, and RX for receive. These are indicator LEDs which blink whenever the UNO board is transmitting or receiving data as shown in Figure 5.40.

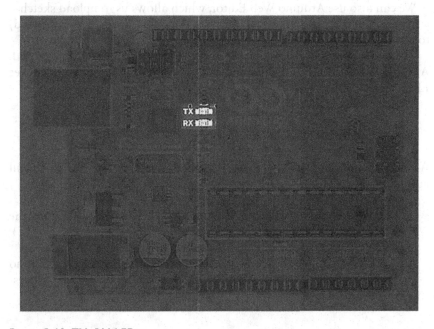

Figure 5.40 TX–RX LEDs.

5.1.7.10 Memory

The memory structure is shown in Figure 5.41:

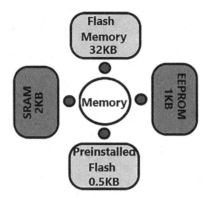

Figure 5.41 Memory.

The preinstalled flash has a bootloader, which takes the memory of 0.5 Kb. Here, SRAM stands for Static Random Access Memory, and EEPROM stands for Electrically Erasable Programmable Read-Only Memory.

5.1.7.11 How to get started with Arduino UNO

We can program the Arduino UNO using the Arduino IDE. The Arduino IDE is the Integral Development program, which is common to all the boards.

We can also use Arduino Web Editor, which allows us to upload sketches and write the code from our web browser (Google Chrome recommended) to any Arduino board. It is an online platform.

The USB connection is essential to connect the computer with the board. After the connection, the PWR pins will light in green. It is a green power LED.

The steps to get started with Arduino UNO are listed below:

Install the **drivers** of the board.

As soon we connect the board to the computer, Windows from XP to 10 will automatically install the board drivers.

- But, if you have expanded or downloaded the zip package, follow the below steps:
- Click on Start->Control Panel->System and Security.
- Click on System->Device Manager->Ports (COM & LPT)->Arduino UNO (COMxx). If the COM &LPT is absent, look Other Devices->Unknown Device.

- Right-click to Arduino UNO (COMxx)->Update Driver Software->Browse my computer for driver software.
- Select the file "inf" to navigate else, select "ArduinoUNO.inf".
- Installation Finished.
- Open the code or sketch written in the Arduino software.
- Select the type of board.

Click on 'Tools' and select Board, as shown below in Figure 5.42:

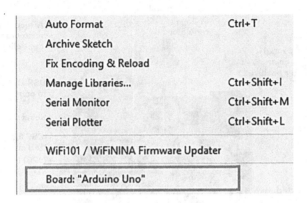

Figure 5.42 How to select board.

Select the port. Click on the Tools -> Port (select the port). The port likely will be COM3 or higher. For example, COM6, etc. The COM1 and COM2ports will not appear, because these two ports are reserved for the hardware serial ports.

Now, upload and run the written code or sketch.

To upload and run, click on the button present on the top panel of the Arduino display, as shown below:

Within the few seconds after the compile and run of code or sketch, the RX and TX lights present on the Arduino board will flash.

The 'Done Uploading' message will appear after the code is successfully uploaded. The message will be visible in the status bar.

5.1.8 Arduino UNO pinout

The Arduino UNO is a standard board of Arduino, which is based on an **ATmega328P** microcontroller. It is easier to use than other types of Arduino boards.

The Arduino UNO board, with the specification of pins, is shown below in Figure 5.43:

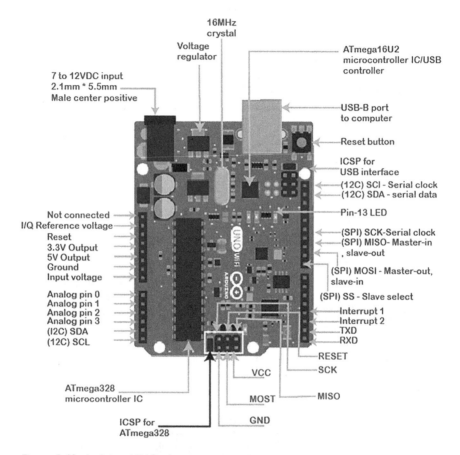

Figure 5.43 Arduino UNO pins.

Let's discuss each pin in detail.

- **ATmega328 Microcontroller**
  This is a single chip microcontroller of the ATmel family. The processor core inside it is an 8-bit one. It is a low-cost, low-powered, and simple microcontroller. The Arduino UNO and Nano models are based on the ATmega328 Microcontroller.

- **Voltage Regulator**
  The voltage regulator converts the input voltage to 5 V. The primary function of the voltage regulator is to regulate the voltage level in the Arduino board. For any changes in the input voltage of the regulator, the output voltage is constant and steady.
- **GND** Ground pins. The ground pins are used to ground the circuit.
- **TXD and RXD**
  TXD and RXD pins are used for serial communication. The TXD is used for transmitting the data, and RXD is used for receiving the data. It also represents the successful flow of data.
- **USB Interface**
  The USB Interface is used to plug in the USB cable. It allows the board to connect to the computer. It is essential for the programming of the Arduino UNO board.
- **RESET**
  It is used to add a Reset button to the connection.
- **SCK**
  It stands for **Serial Clock**. These are the clock pulses, which are used to synchronize the transmission of data.
- **MISO**
  It stands for **Master Input/Slave Output**. The save line in the MISO pin is used to send the data to the master.
- **VCC**
  This is the modulated DC supply voltage, which is used to regulate the IC's used in the connection. It is also called the primary voltage for IC's present on the Arduino board. The Vcc voltage value can be either negative or positive with respect to the GND pin.
- **Crystal Oscillator** The Crystal oscillator has a frequency of 16 MHz, which makes the Arduino UNO a powerful board.
- **ICSP**
  This stands for **In-Circuit Serial Programming**. The users can program the Arduino board's firmware using the ICSP pins.

The program or firmware with the advanced functionalities is received by the microcontroller with the help of the ICSP header.

The ICSP header consists of 6 pins.

The structure of the ICSP header is shown below in Figure 5.44:

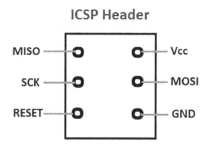

Figure 5.44 ICSP header.

This is the top view of the ICSP header.

- **SDA**
  This stands for **Serial Data**. It is a line used by the slave and master to send and receive data. It is called as a **data line,** while SCL is called a clock line.
- **SCL**
  This stands for **Serial Clock**. It is defined as the line that carries the clock data. It is used to synchronize the transfer of data between the two devices. The Serial Clock is generated by the device and it is called the master.
- **SPI**
  This stands for **Serial Peripheral Interface**. It is popularly used by the microcontrollers to communicate with one or more peripheral devices quickly. It uses conductors for data receiving, data sending, synchronization, and device selection (for communication).
- **MOSI**
  This stands for Master Output/Slave Input.
  The MOSI and SCK are driven by the master.
- **SS**
  This stands for **Slave Select**. It is the Slave Select line, which is used by the master. It acts as the enable line.
- **I2C**
  This is the two-wire serial communication protocol. It stands for Inter Integrated Circuits. The I2C is a serial communication protocol that uses SCL (Serial Clock) and SDA (Serial Data) to receive and send data between two devices.
  3.3 V and 5 V are the operating voltages of the board.

5.1.8.1 *Arduino coding basics*

Arduino IDE (the Integrated Development Environment) allows us to draw the sketch and upload it to the various Arduino boards using code. The code is written in a simple programming language similar to C and C++. The initial step to start with Arduino is the IDE download and installation.

To start with Arduino programming:

5.1.8.1.1 Brackets

There are two types of brackets used in the Arduino coding, which are listed below:

- Parentheses ( )
- Curly Brackets { }

   Parentheses ( ): The parentheses brackets are the group of the arguments, such as method, function, or a code statement. These are also used to group the math equations.

   Curly Brackets { }: The statements in the code are enclosed in the curly brackets. We always require closed curly brackets to match the open curly bracket in the code or sketch.

   Open curly **bracket-** '{ '

   Closed **curly** bracket – ' } '

5.1.8.1.2 Line comment

There are two types of line comments, which are listed below:

- Single line comment
- Multi-line comment

// Single line comment

The text that is written after the two forward slashes are considered as a single line comment. The compiler ignores the code written after the two forward slashes. The comment will not be displayed in the output. Such text is specified for a better understanding of the code or for the explanation of any code statement.

The // (two forward slashes) are also used to ignore some extra lines of code without deleting it.

/ * Multi-line comment */

The multi-line comment is written to group the information for clear understanding. It starts with the single forward slash and an asterisk symbol (/ *). It also ends with the/ *. It is commonly used to write the larger text. It is a comment, which is also ignored by the compiler.

5.1.8.1.3 Coding screen

The coding screen is divided into two blocks. The **setup** is considered as the preparation block, while the **loop** is considered as the execution block. It is shown below in Figure 5.45:

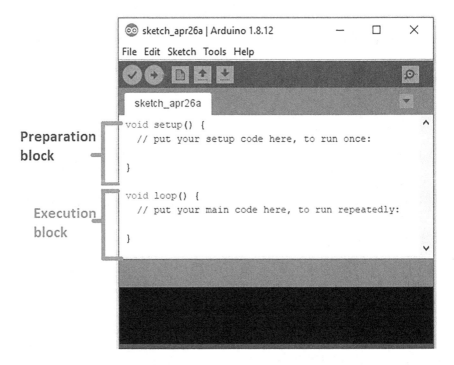

Figure 5.45 Arduino sketch.

The set of statements in the setup and loop blocks are enclosed with the curly brackets. We can write multiple statements depending on the coding requirements for a particular project.

For example:

```
void setup()
{
Codingstatement1;
Codingstatement2;
.
.
.
Codingstatementn;
}
```

```
void loop()
{
Codingstatement1;
Codingstatement2;
.
.
.
Codingstatementn;
}
```

5.2 WHAT IS SETUP? WHAT TYPE OF CODE IS WRITTEN IN THE SETUP BLOCK?

It contains an initial part of the code to be executed. The pin modes, libraries, variables, etc., are initialized in the setup section. It is executed only once during the uploading of the program and after reset or power up of the Arduino board.

Zero setup () resides at the top of each sketch. As soon as the program starts running, the code inside the curly bracket is executed in the setup and it executes only once.

5.3 WHAT IS LOOP? WHAT TYPE OF CODE IS WRITTEN IN THE LOOP BLOCK?

The loop contains statements that are executed repeatedly. The section of code inside the curly brackets is repeated depending on the value of variables.

Time in Arduino: The time in Arduino programming is measured in a millisecond.

Where, 1 sec = 1000 milliseconds

We can adjust the timing according to the milliseconds.

For example, for a 5-second delay, the time displayed will be 5000 milliseconds.

Example: Let's consider a simple LED blink example.

The steps to open such examples are:

1. Click on the **File** button, which is present on the menu bar.
2. Click on the **Examples**.
3. Click on the **Basics** option and click on the **Blink** button.

The example will reopen in a new window, as shown below in Figure 5.46:

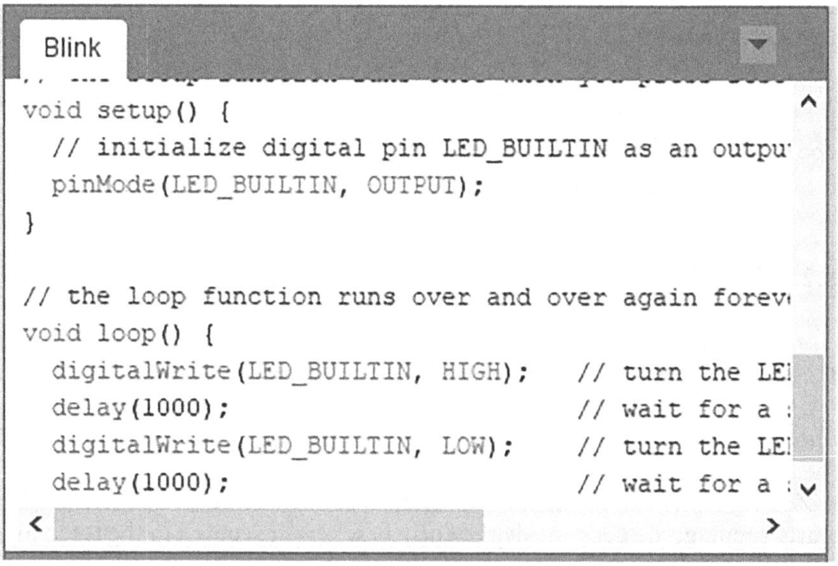

```
Blink                                                        ▼

void setup() {
    // initialize digital pin LED_BUILTIN as an outpu'
    pinMode(LED_BUILTIN, OUTPUT);
}

// the loop function runs over and over again forev
void loop() {
    digitalWrite(LED_BUILTIN, HIGH);    // turn the LE
    delay(1000);                        // wait for a
    digitalWrite(LED_BUILTIN, LOW);     // turn the LE
    delay(1000);                        // wait for a
```

Figure 5.46 LED blink example.

- The void setup () would include pinMode as the main function. pinMode ()

The specific pin number is set as the INPUT or OUTPUT in the pinMode () function.
The Syntax is: **pinMode (pin, mode)**
Where,

pin: It is the pin number. We can select the pin number according to the requirements.
Mode: We can set the mode as INPUT or OUTPUT according to the corresponding pin number.

Let' understand the pinMode with an example.

Example: We want to set the 12 pin number as the output pin.
Code:
pinMode(12,OUTPUT);

5.4 WHY IS IT RECOMMENDED TO SET THE MODE OF PINS AS OUTPUT?

The OUTPUT mode of a specific pin number provides a considerable amount of current to other circuits, which is enough to run a sensor or to light the LED brightly. The output state of a pin is considered as the low-impedance state. The high current and short circuit of a pin can damage the ATmel chip. So, it is recommended to set the mode as OUTPUT.

5.4.1 Can we set the pinMode as INPUT?

The digitalWrite () will disable the LOW during the INPUT mode. The output pin will be considered as HIGH. We can use the INPUT mode to use the external pull-down resistor. We are required to set the pinMode as INPUT_PULLUP. It is used to reverse the nature of the INPUT mode. The sufficient amount of current is provided by the pull-up mode to dimly light an LED, which is connected to the pin in the INPUT mode. If the LED is working dimly, it means this condition is working out. Due to this, it is recommended to set the pin in OUTPUT mode.

- The void loop () would include **digitalWrite( )** and **delay( )** as the main function.

digitalWrite( )

The digitalWrite ( ) function is used to set the value of a pin as HIGH or LOW.

Where, **HIGH:** It sets the value of the voltage. For the 5 V board, it will set the value of 5 V, while for 3.3 V; it will set the value of 3.3 V. **LOW:** It sets the value = 0 (GND).If we do not set the pinMode as OUTPUT, the LED may light dimly.

The syntax is: **digitalWrite(pin, value HIGH/LOW)**

pin: We can specify the pin number or the declared variable.

Let's understand with an example.

Example:

digitalWrite(13,HIGH);
digitalWrite(13,LOW);
The HIGH will ON the LED and LOW will OFF the LED connected to pin number 13.

5.5 WHAT IS THE DIFFERENCE BETWEEN DigitalRead () AND DigitalWrite ()?

The digitalRead () function will read the HIGH/LOW value from the digital pin, and the digitalWrite () function is used to set the HIGH/LOW value of the digital pin.

delay ( )

The delay () function is a blocking function to pause a program from doing a task during the specified duration in milliseconds.

For example, - delay (2000)
Where, 1 sec = 1000 milliseconds
Hence, it will provide a delay of 2 seconds.
Code:

```
      digitalWrite(13,HIGH);
   1. delay(2000);
   2. digitalWrite(13,LOW);
   3. delay(1000);
```

Here, the LED connected to pin number 13 will be ON for 2 seconds and OFF for 1 second. The task will repeatedly execute as it is in the void loop ().We can set the duration according to our choice or project requirements.

> **Example:** To light the LED connected to pin number 13. We want to
> ON the LED for 4 seconds and OFF the LED for 1.5 seconds.

Code:

```
voidsetup()
{
pinMode(13,OUTPUT);//to set the OUTPUT mode of pin number 13.
}
voidloop()
{
digitalWrite(13,HIGH);
delay(4000);//4 seconds = 4 × 1000 milliseconds
digitalWrite(13,LOW);
delay(1500);//1.5 seconds = 1.5 × 1000 milliseconds
}
```

5.5.1 Arduino syntax and program flow

Syntax:
Syntax in Arduino signifies the rules need to be followed for the successful uploading of the Arduino program to the board. The syntax of Arduino is similar to the grammar in English. It means that the rules must be followed in order to compile and run our code successfully. If we break those rules, our computer program may compile and run, but with some bugs.

Let's understand with an example. As we open the Arduino IDE, the display will look like Figure 5.47:

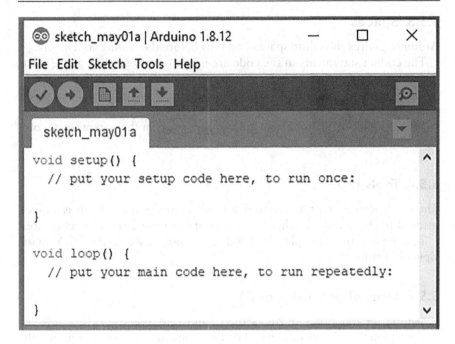

Figure 5.47 Arduino structure.

The two functions that encapsulate the pieces of code in the Arduino program are shown below:

1. **void setup ( )**
2. **void loop ( )**

5.5.2 Functions

- The functions in Arduino combine many pieces of lines of code into one.
- The functions usually return a value after finishing execution. But here, the function does not return any value due to the presence of void.
- The setup and loop function have **void** keyword present in front of their function name.
- The multiple lines of code that a function encapsulates are written inside curly brackets.
- Every closing curly bracket '}' must match the opening curly bracket '{' in the code.
- We can also write our own functions, which will be discussed later in this tutorial.

5.5.3 Spaces

Arduino ignores the white spaces and tabs before the coding statements.

The coding statements in the code are intent (empty spacing at the starting) for the easy reading.

In the function definition, loop, and conditional statements, 1 intent = 2 spaces.

The compiler of Arduino also ignores the spaces in the parentheses, commas, blank lines, etc.

5.5.4 Tools tab

The verify icon present on the tool tab only compiles the code. It is a quick method to check that whether the syntax of our program is correct or not.

To compile, run, and upload the code to the board, we need to click on the Upload button.

5.5.5 Uses of parentheses ( )

It denotes the function like void setup ( ) and void loop ( ).

The parameter's inputs to the function are enclosed within the parentheses.

It is also used to change the order of operations in mathematical operations.

5.5.6 Semicolon ;

It is the statement terminator in the C as well as C++.

A statement is a command given to the Arduino, which instructs it to take some kind of action. Hence, the terminator is essential to signify the end of a statement.

We can write one or more statements in a single line, but with semicolon indicating the end of each statement.

The compiler will indicate an error if a semicolon is absent in any of the statements.

It is recommended to write each statement with semicolon in a different line, which makes the code easier to read.

We are not required to place a semicolon after the curly braces of the setup and loop function.

Arduino processes each statement sequentially. It executes one statement at a time before moving to the next statement.

5.5.7 Program flow

The program flow in Arduino is similar to the flowcharts. It represents the execution of a program in order.

We recommend to draw the flowchart before writing the code. It helps us to understand the concept of code, which makes it the coding simpler and easier.

5.5.8 Flow charts

A flowchart uses shapes and arrows to represent the information or sequence of actions.

An oval ellipse shows the Start of the sequence, and a square shows the action or processes that need to be performed.

The Arduino coding process in the form of the flowchart is shown below:

Setup ( ): Acts as an entry point

Here, the processor enters our code, and the execution of code begins. After the setup, the execution of the statement in the loop begins.

loop ( ) : runs over and over again

The example of the flowchart in Arduino is shown below in Figure 5.48:

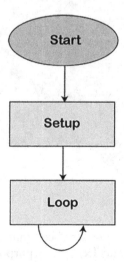

Figure 5.48 Arduino flowchart.

5.5.9 Arduino Serial |Serial.begin()

Serial Communication: The serial communication is a simple scheme that uses the **UART** (Universal Asynchronous Receiver/Transmitter) on the Microcontroller. It uses:

5 V for logic 1 (high)
0 V for logic 0 (low)

For a 3.3 V board, it uses:

3 V for logic 1 (high)
0 V for logic 0 (low)

Every message sent on the UART is in the form of 8 bits or 1 byte, where1 **byte = 8 bits.**

The messages sent to the computer from Arduino are **sent from PIN 1 of the** Arduino board, **called Tx (Transmitter).** The messages being sent to the Arduino from the computer are **received on PIN 0, called Rx(Receiver).**

These two pins on the Arduino UNO board look like the below image in Figure 5.49:

Figure 5.49 UART.

When we initialize the pins for serial communication in our code, we cannot use these two pins (Rx and Tx) for any purpose. The Tx and Rx pins are also connected directly to the computer.

The pins are connected to the serial Tx and Rx chip, which acts as a serial to USB translator. It acts as a medium for the computer to talk to the Microcontroller.

The chip on the board looks like the below image in Figure 5.50:

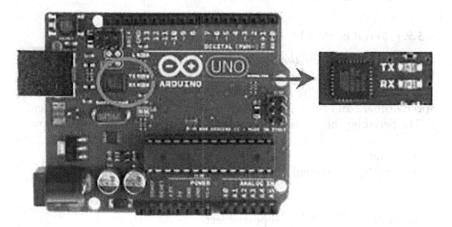

Figure 5.50 Arduino.

The object can include any number of data members (information) and member functions (to call actions).

The Serial.begin( ) is a part of the serial object in the Arduino. It tells the serial object to perform initialization steps to send and receive data on the Rx and Tx (pins 1 and 0).

Let's discuss Serial.begin( ) in detail.

Arduino Mega has four serial ports. The Tx pins on the Mega board are listed below:

- 1 (TX)
- 18 (TX)
- 16 (TX)
- 14 (TX)

The Rx pins on the Mega port are listed below:

- 0 (RX)
- 19 (RX)
- 17 (RX)
- 15 (RX)

The communication with the Tx and Rx pins would cause interference and failed uploads to the particular board.

If we require a serial port for communication, we need to use a **USB-to-serial adapter**. It is a mini USB connector, which converts the USB connection to the Serial RX and TX. We can directly connect the adapter to the board.

There are five pins present on the USB-to serial adapter, including RX, TX, reset button, and GND (Ground).

5.5.9.1 Serial.begin ( )

The serial.begin( ) *sets the baud rate for serial data communication.* The **baud** rate signifies the data rate in bits per second.

The default baud rate in Arduino is **9600 bps (bits per second)**. We can specify other baud rates as well, such as 4800, 14400, 38400, 28800, etc.

The Serial.begin( ) is declared in two formats, which are shown below:

- begin(speed)
- begin(speed, config)

Where,

    serial: It signifies the serial port object.

    speed: It signifies the baud rate or bps (bits per second) rate. It allows long data types.

    config: It sets the stop, parity, and data bits.

Example 1:

```
1. voidsetup()
2. {
3. Serial.begin(4800);
4. }
5. voidloop()
6. {
7. }
```

The serial.begin (4800) open the serial port and set the bits per rate to 4800. The messages in Arduino are interchanged with the serial monitor at a rate of 4800 bits per second.

5.5.9.2 Arduino serial.print ( )

The serial.print ( ) in Arduino prints the data to the serial port. The printed data is stored in the ASCII (American Standard Code for Information Interchange) format, which is a human-readable text.

Each digit of a number is printed using the ASCII characters.

The printed data will be visible in the **serial monitor**, which is present on the right corner on the toolbar.

The Serial.print( ) is declared in two formats, which are shown below:

- print(value)
- print(value, format)

Note: In Serial.print(), S must be written in uppercase.

Where,

serial: It signifies the serial port object.

print: The print ( ) returns the specified number of bytes written.

value: It signifies the value to print, which includes any data type value.

format: It consists of number base, such as OCT (Octal), BIN (Binary), HEX (Hexadecimal), etc. for the integral data types. It also specifies the number of decimal places.

5.5.9.3 Serial.print(value)

The serial.print ( ) accepts the number using the ASCII character per digit and value up to two decimal places for floating point numbers.

Example 1:

1. Serial.print(15.452732)

   Output:
   15.45
   It sends bytes to the printer as a single character. In Arduino, the strings and characters using the Serial.print( ) are sent as it is.

Example 2:

1. Serial.print("HelloArduino")

   Output:
   "Hello Arduino"
   Serial.print(value, format)

   It specifies the base format and gives the output according to the specified format. It includes the formats Octal -OCT (base 8), Binary-BIN (base 2), Decimal-DEC (base 10), and Hexadecimal-HEX (base 16). Let's understand by few examples.

Example 1:

1. Serial.print(25,BIN)

Output:
11001
It converts the decimal number 25 to binary number 11001.

Example 2:

1. Serial.print(58,HEX)

Output:
3 A
It converts the decimal number 58 to hexadecimal number 3 A.
Serial.println ( )
The Serial.println ( ) means print line, which sends the string followed by the carriage return ('\r' or ASCII 13) and newline ('\n' or ASCII 10) characters. It has a similar effect as pressing the Enter or Return key on the keyboard when typing with the Text Editor.
The Serial.println( ) is also declared in two formats, which are shown below:

- println(value)
- println(value, format)

5.6 WHAT IS THE DIFFERENCE BETWEEN Serial.Print( ) AND Serial.Println( )?

The text written inside the open and closed parentheses in the Serial.println( ) moves in a new line. With the help of Serial.print( ) and Serial.println( ), we can figure the order and execution of certain things in our code.
Let's understand with an example:
Consider the below code.

```
1. voidsetup()
2. {
3. Serial.begin(4800);
4. }
5. voidloop()
6. {
7. Serial.print("Hello");
8. delay(1000);
9. Serial.println("Arduino");//It will print Arduino
followed by a new line.
10. delay(1500);//delay of 1.5 seconds between each printed
line.
11. }
```

Click on the **Upload** button->**Serial monitor** for the output.

In the output, the word **Hello** will appear followed by the word **Arduino**1 second later. After 1.5 second, another line will be printed.

Output

Hello Arduino

Hello Arduino//The next line will be printed after the specified duration.

The output will be printed repeatedly.

Program 2: Turn on LED

```
int LED = 6;
                // LED connected to pin 6
 void setup ()
{ pinMode(LED, OUTPUT); // set the digital pin as output
}
void loop () {
digitalWrite(LED,HIGH); // turn on led
}
```

Program 3: Blinking LED

```
int LED = 6;
                // LED connected to pin 6
 void setup ()
{ pinMode(LED, OUTPUT); // set the digital pin as output
}
void loop () {
digitalWrite(LED,HIGH); // turn on led
delay(500); // delay for 500 ms, 1 second=1000 milliseconds
digitalWrite(LED,LOW);   // turn off led
delay(500); // delay for 500 ms
}
```

5.6.1 Data types

Data types in C refer to an extensive system used for declaring variables or functions of different types. The type of a variable determines how much space it occupies in the storage and how the bit pattern stored is interpreted. The Arduino environment is really just C++ with library support and built-in assumptions about the target environment to simplify the coding process. C++ defines a number of different data types; here we'll talk only about those used in Arduino with an emphasis on traps awaiting the unwary Arduino programmer.

Below is a list of the data types commonly seen in Arduino, with the memory size of each in parentheses after the type name. Note: **signed** variables allow both positive and negative numbers, while **unsigned** variables allow only positive values.

- **Boolean**(8 bit) – simple logical true/false
- **byte**(8 bit) – unsigned number from 0-255
- **char**(8 bit) – signed number from -128 to 127. The compiler will attempt to interpret this data type as a character in some circumstances, which may yield unexpected results
- **unsigned char** (8 bit) – same as 'byte'; if this is what you're after, you should use 'byte' instead, for reasons of clarity
- **word**(16 bit) – unsigned number from 0-65535
- **unsigned int**(16 bit) – the same as 'word'. Use 'word' instead for clarity and brevity
- **int**(16 bit) – signed number from -32768 to 32767. This is most commonly what you see used for general purpose variables in Arduino example code provided with the IDE
- **unsigned long**(32 bit) – unsigned number from 0-4,294,967,295. The most common usage of this is to store the result of the millis()function, which returns the number of milliseconds the current code has been running
- **long**(32 bit) – signed number from -2,147,483,648 to 2,147,483,647
- **float**(32 bit) – signed number from -3.4028235E38 to 3.4028235E38. Floating point on the Arduino is not native; the compiler has to jump through hoops to make it work.
- The following table provides all the data types that you will use during Arduino programming:

void	Boolean	char	Unsigned char	byte	int	Unsigned int	word
long	Unsigned long	short	float	double	array	String-char array	String-object

5.6.1.1 Void

The void keyword is used only in function declarations. It indicates that the function is expected to return no information to the function from which it was called.

Example

```
Void Loop ( ) {
// rest of the code
}
```

5.6.1.2 Boolean

A Boolean holds one of two values, true or false. Each Boolean variable occupies one byte of memory.

Example

> boolean val = false; // declaration of variable with type boolean and initialize it with false
>
> boolean state = true; // declaration of variable with type boolean and initialize it with true

5.6.1.3 Char

A data type that takes up one byte of memory that stores a character value. Character literals are written in single quotes like this: 'A' and for multiple characters, strings use double quotes: "ABC".

However, characters are stored as numbers. You can see the specific encoding in the ASCII chart. This means that it is possible to do arithmetic operations on characters, in which the ASCII value of the character is used. For example, 'A' + 1 have the value 66, since the ASCII value of the capital letter A is 65.

Example

> Char chr_a = 'a';//declaration of variable with type char and initialize it with character a
>
> Char chr_c = 97;//declaration of variable with type char and initialize it with character 97

5.6.1.4 Unsigned char

Unsigned char is an unsigned data type that occupies one byte of memory. The unsigned char data type encodes numbers from 0 to 255.

Example

> Unsigned Char chr_y = 121; // declaration of variable with type Unsigned char and initialize it with character y

5.6.1.5 Byte

A byte stores an 8-bit unsigned number, from 0 to 255.

Example

> byte m = 25;//declaration of variable with type byte and initialize it with 25

5.6.1.6 Int

Integers are the primary data-type for number storage. int stores a 16-bit (2-byte) value. This yields a range of -32,768 to 32,767 (minimum value of -2^{15} and a maximum value of $(2^{15}) - 1$).

The **int** size varies from board to board. On the Arduino Due, for example, an **int** stores a 32-bit (4-byte) value. This yields a range of -2,147,483,648 to 2,147,483,647 (minimum value of -2^{31} and a maximum value of $(2^{31}) - 1$).

Example

> int counter = 32;// declaration of variable with type int and initialize it with 32

5.6.1.7 Unsigned int

Unsigned ints (unsigned integers) are the same as int in the way that they store a 2 byte value. Instead of storing negative numbers, however, they only store positive values, yielding a useful range of 0 to 65,535 (2^{16} - 1). The Due stores a 4 byte (32-bit) value, ranging from 0 to 4,294,967,295 (2^{32} - 1).

Example

> Unsigned int counter = 60; // declaration of variable with
> type unsigned int and initialize it with 60

5.6.1.8 Word

On the Uno and other ATMEGA-based boards, a word stores a 16-bit unsigned number. On the Due and Zero, it stores a 32-bit unsigned number.

Example

> word w = 1000;//declaration of variable with type word and initialize it with 1000

5.6.1.9 Long

Long variables are extended size variables for number storage, and store 32 bits (4 bytes), from -2,147,483,648 to 2,147,483,647.

Example

> Long velocity = 102346;//declaration of variable with type Long and initialize it with 102346

5.6.1.10 Unsigned long

Unsigned long variables are extended size variables for number storage and store 32 bits (4 bytes). Unlike standard longs, unsigned longs will not store negative numbers, making their range from 0 to 4,294,967,295 ($2^{32} - 1$).

Example

Unsigned Long velocity = 101006;// declaration of variable with type Unsigned Long and initialize it with 101006

5.6.1.11 Short

A short is a 16-bit data-type. On all Arduinos (ATMega and ARM-based), a short stores a 16-bit (2-byte) value. This yields a range of -32,768 to 32,767 (minimum value of -2^{15} and a maximum value of $(2^{15}) - 1$).

Example

short val = 13;//declaration of variable with type short and initialize it with 13

5.6.1.12 Float

Data type for floating-point number is a number that has a decimal point. Floating-point numbers are often used to approximate the analog and continuous values because they have greater resolution than integers.

Floating-point numbers can be as large as 3.4028235E+38 and as low as -3.4028235E+38. They are stored as 32 bits (4 bytes) of information.

Example

float num = 1.352;//declaration of variable with type float and initialize it with 1.352

5.6.1.13 Double

On the Uno and other ATMEGA based boards, Double precision floating-point number occupies four bytes. That is, the double implementation is exactly the same as the float, with no gain in precision. On the Arduino Due, doubles have 8-byte (64 bit) precision.

Example

double num = 45.352;// declaration of variable with type double and initialize it with 45.352

Chapter 6

Introduction to Arduino

Arduino is an open-source electronics platform based on easy-to-use hardware and software. Arduino boards are able to read inputs: light on a sensor, a finger on a button, or a Twitter message and turn it into an output: activating a motor, turning on an LED, publishing something online. You can tell your board what to do by sending a set of instructions to the microcontroller on the board. To do so, you use the Arduino programming language (based on Wiring) and the Arduino Software (IDE) based on Processing.

Over the years Arduino has been the brain of thousands of projects, from everyday objects to complex scientific instruments. A worldwide community of makers, students, hobbyists, artists, programmers and professionals, has gathered around this open-source platform. Their contributions have added up to an incredible amount of accessible knowledge that can be of great help to novices and experts alike.

Arduino was born at the Ivrea Interaction Design Institute as an easy tool for fast prototyping, aimed at students without a background in electronics and programming. As soon as it reached a wider community, the Arduino board started changing to adapt to new needs and challenges, differentiating its offer from simple 8-bit boards to products for IoT applications, and wearable, 3D printing and embedded environments.

6.1 WHY ARDUINO?

Thanks to its simple and accessible user experience, Arduino has been used in thousands of different projects and applications. The Arduino software is easy to use for beginners, yet flexible enough for advanced users. It runs on Mac, Windows, and Linux. Teachers and students use it to build low-cost scientific instruments, to prove chemistry and physics principles or to get started with programming and robotics. Designers and architects build

DOI: 10.1201/9781003307488-6

interactive prototypes, musicians and artists use it for installations and to experiment with new musical instruments. Makers, of course, use it to build many of the projects exhibited at the Maker Faire, for example. Arduino is a key tool to learn new things. Anyone, including children, hobbyists, artists, and programmers, can begin to tinker just following the step-by-step instructions of a kit or sharing ideas online with other members of the Arduino community.

There are many other microcontrollers and microcontroller platforms available for physical computing. Parallax Basic Stamp, Netmedia's BX-24, Phidgets, MIT's Handyboard, and many others offer similar functionality. All of these tools take the messy details of microcontroller programming and wrap it up in an easy-to-use package. Arduino also simplifies the process of working with microcontrollers, but it offers some advantages for teachers, students, and interested amateurs over other systems:

- Inexpensive: Arduino boards are relatively inexpensive compared to other microcontroller platforms. The least expensive version of the Arduino module can be assembled by hand and even the pre-assembled Arduino modules cost less than $50.
- Cross-platform: The Arduino Software (IDE) runs on Windows, Macintosh OSX, and Linux operating systems. Most microcontroller systems are limited to Windows.
- Simple, clear programming environment: The Arduino Software (IDE) is easy to use for beginners, yet also flexible enough for advanced users to take advantage of. For teachers, it's conveniently based on the Processing programming environment, so students learning to program in that environment will be familiar with how the Arduino IDE works.
- Open source and extensible software: The Arduino software is published as open-source tools, available for extension by experienced programmers. The language can be expanded through C++ libraries and people wanting to understand the technical details can make the leap from Arduino to the C programming language on which it's based. Similarly, if you want to you can add C code directly into your Arduino programs.
- Open source and extensible hardware: The plans of the Arduino boards are published under a Creative Commons license, so experienced circuit designers can make their own version of the module, extending it and improving it. Even relatively inexperienced users can build the breadboard version of the module in order to understand how it works and save money.

6.2 WHAT MAKES UP AN ARDUINO?

Arduinos contain a number of different parts and interfaces together on a single circuit board. The design has changed through the years, and some variations also include other parts. But on a basic board, you're likely to find the following pieces:

- A number of pins, which are used to connect with various components you might want to use with the Arduino. These pins come in two varieties:
  - Digital pins: This can read and write a single state, on or off. Most Arduinos have 14 digital I/O pins.
  - Analog pins: This can read a range of values, and are useful for more fine-grained control. Most Arduinos have six of these analog pins.

These pins are arranged in a specific pattern, so that if you buy an add-on board designed to fit into them, typically called a "shield," it should fit easily into most Arduino-compatible devices.

- A power connector, which provides power to both the device itself and provides a low voltage which can power connected components like LEDs and various sensors, provided their power needs are reasonably low. The power connector can connect to either an AC adapter or a small battery.
- A microcontroller, the primary chip, which allows you to program the Arduino in order for it to be able to execute commands and make decisions based on various input. The exact chip varies depending on what type of Arduino you buy, but they are generally Atmel controllers, usually an ATmega8, ATmega168, ATmega328, ATmega1280 or ATmega2560. The differences between these chips are subtle but the biggest difference a beginner will notice is the different amounts of onboard memory.
- A serial connector, which on most newer boards is implemented through a standard USB port. This connector allows you to communicate to the board from your computer, as well as load new programs onto the device. Often, Arduinos can also be powered through the USB port, removing the need for a separate power connection.
- A variety of other small components, like an oscillator and/or a voltage regulator, which provide important capabilities to the board, although you typically don't interact with these directly; just know that they are there.

6.3 WHAT DOES IT DO?

The Arduino hardware and software was designed for artists, designers, hobbyists, hackers, newbies, as shown in Figure 6.1 and anyone interested in creating interactive objects or environments. Arduino can interact with buttons, LEDs, motors, speakers, GPS units, cameras, the Internet, and even your smartphone or your TV. This flexibility combined with the fact that the Arduino software is free, the hardware boards are pretty cheap, and both the software and hardware are easy to learn has led to a large community of users who have contributed code and released instructions for a huge variety of Arduino-based projects.

For everything from robots and a heating pad and hand warming blanket to honest fortune-telling machines and even a Dungeons and Dragons dice-throwing gauntlet, the Arduino can be used as the brains behind almost any electronics project.

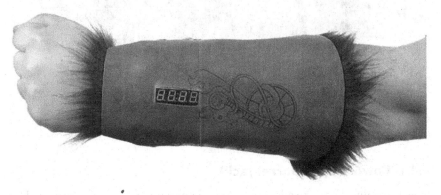

Figure 6.1 Arduino (Wear your nerd cred on your sleepy arm).

6.4 WHAT'S ON THE BOARD?

There are many varieties of Arduino boards that can be used for different purposes as shown in Figure 6.2. Some boards look a bit different from the one below, but most Arduinos have the majority of these components in common:

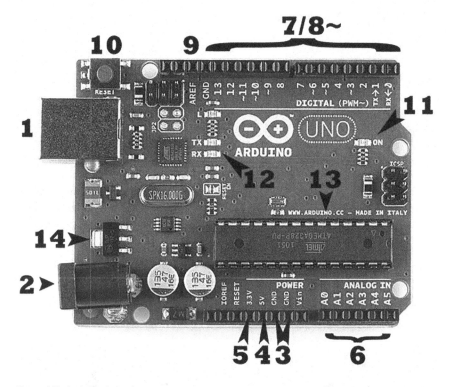

Figure 6.2 Arduino board.

6.4.1 Power (USB/Barrel Jack)

Every Arduino board needs a way to be connected to à power source. The Arduino UNO can be powered from a USB cable coming from your computer or a wall power supply that is terminated in a barrel jack. In Figure 6.2, the USB connection is labeled (1) and the barrel jack is labeled (2). The USB connection is also how you will load code onto your Arduino board.

Do NOT use a power supply greater than 20 Volts as you will overpower (and thereby destroy) your Arduino. The recommended voltage for most Arduino models is between 6 and 12 Volts.

6.4.2 Pins (5V, 3.3V, GND, Analog, Digital, PWM, AREF)

The pins on your Arduino are the places where you connect wires to construct a circuit (probably in conjunction with a breadboard and some wire). They usually have black plastic 'headers' that allow you to just plug a wire right into the board. The Arduino has several different kinds of pins, each of which is labeled on the board and used for different functions.

- **GND (3)**: Short for 'Ground'. There are several GND pins on the Arduino, any of which can be used to ground your circuit.
- **5V (4) & 3.3V (5)**: As you might guess, the 5V pin supplies 5 volts of power and the 3.3V pin supplies 3.3 volts of power. Most of the simple components used with the Arduino run happily off either 5 or 3.3 volts.
- **Analog (6)**: The area of pins under the 'Analog In' label (A0 through A5 on the UNO) is Analog In pins. These pins can read the signal from an analog sensor (like a temperature sensor) and convert it into a digital value that we can read.
- **Digital (7)**: Across from the analog pins are the digital pins (0 through 13 on the UNO). These pins can be used for both digital input (like telling if a button is pushed) and digital output (like powering an LED).
- **PWM (8)**: You may have noticed the tilde (~) next to some of the digital pins (3, 5, 6, 9, 10, and 11 on the UNO). These pins act as normal digital pins but can also be used for something called Pulse-Width Modulation (PWM). Think of these pins as being able to simulate analog output (like fading an LED in and out).
- **AREF (9)**: Stands for Analog Reference. Most of the time you can leave this pin alone. It is sometimes used to set an external reference voltage (between 0 and 5 Volts) as the upper limit for the analog input pins.

6.4.2.1 Reset Button

Just like the original Nintendo, the Arduino has a reset button (10). Pushing it will temporarily connect the reset pin to ground and restart any code that is loaded on the Arduino. This can be very useful if your code doesn't repeat, but you want to test it multiple times. Unlike the original Nintendo, however, blowing on the Arduino doesn't usually fix any problems!

6.4.2.2 Power LED Indicator

Just beneath and to the right of the word "UNO" on your circuit board, there's a tiny LED next to the word 'ON' (11). This LED should light up whenever you plug your Arduino into a power source. If this light doesn't turn on, there's a good chance something is wrong. Time to re-check your circuit!

6.4.2.3 TX RX LEDs

TX is short for transmit; RX is short for receive. These markings appear fairly often in electronics to indicate the pins responsible for serial communication.

In our case, there are two places on the Arduino UNO, where TX and RX appear once by digital pins 0 and 1 and a second time next to the TX and RX indicator LEDs (12). These LEDs will give us some nice visual indications whenever our Arduino is receiving or transmitting data (like when we're loading a new program onto the board).

6.4.2.4 Main IC

The black thing with all the metal legs is an IC, or Integrated Circuit (13). Think of it as the brains of our Arduino. The main IC on the Arduino is slightly different from board type to board type, but is usually from the ATmega line of IC's from the ATMEL company. This can be important, as you may need to know the IC type (along with your board type) before loading up a new program from the Arduino software. This information can usually be found in writing on the top side of the IC. If you want to know more about the difference between various IC's, reading the datasheets is often a good idea.

6.4.2.5 Voltage Regulator

The voltage regulator (14) is not actually something you can (or should) interact with on the Arduino. But it is potentially useful to know that it is there and what it's for. The voltage regulator does exactly what it says – it controls the amount of voltage that is let into the Arduino board. Think of it as a kind of gatekeeper; it will turn away an extra voltage that might harm the circuit. Of course, it has its limits, so don't hook up your Arduino to anything greater than 20 volts.

6.4.2.6 The Arduino Family

Arduino makes several different boards, each with different capabilities. In addition, part of being open-source hardware means that others can modify and produce derivatives of Arduino boards that provide even more form factors and functionality.

6.4.2.6.1 Arduino UNO (R3)

The UNO is a great choice for your first Arduino. It has 14 digital input/output pins (of which 6 can be used as PWM outputs), 6 analog inputs, a USB connection, a power jack, a reset button and more as shown in Figure 6.3. It contains everything needed to support the microcontroller; simply connect it to a computer with a USB cable or power it with an AC-to-DC adapter or battery to get started.

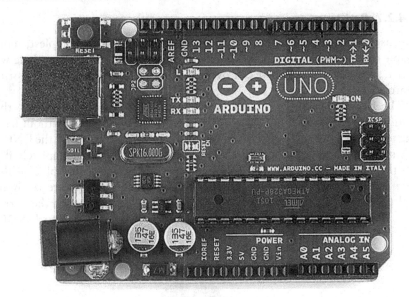

Figure 6.3 Arduino UNO.

6.4.2.6.2 *LilyPad Arduino*

This is the LilyPad Arduino main board! LilyPad is a wearable e-textile technology developed by Leah Buechley and cooperatively designed by Leah and SparkFun. Each LilyPad was creatively designed with large connecting pads and a flat back to allow them to be sewn into clothing with conductive thread. The LilyPad also has its own family of input, output, power, and sensor boards that are also built specifically for e-textiles as shown in Figure 6.4. They're even washable!

Figure 6.4 Lilypad Arduino.

6.4.2.7 RedBoard

We use many Arduinos and we're always looking for the simplest, most stable one. Each board is a bit different and no one board has everything we want, so we decided to make our own version that combines all our favorite features.

The RedBoard can be programmed over a USB Mini-B cable using the Arduino IDE as shown in Figure 6.5. It'll work on Windows 8 without having to change your security settings (we used signed drivers, unlike the UNO). It's more stable due to the USB/FTDI chip we used, plus it's completely flat on the back, making it easier to embed in your projects. Just plug in the board, select "Arduino UNO" from the board menu and you're ready to upload code. You can power the RedBoard over USB or through the barrel jack. The on-board power regulator can handle anything from 7 to 15VDC.

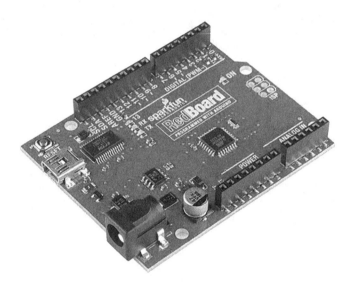

Figure 6.5 RedBoard.

6.4.2.7.1 Arduino Mega (R3)

The Arduino Mega is like the UNO's big brother. It has lots (*54!*) of digital input/output pins (14 can be used as PWM outputs), 16 analog inputs, a USB connection, a power jack, and a reset button. It contains everything needed to support the microcontroller; simply connect it to a computer with a USB cable or power it with an AC-to-DC adapter or battery to get started as shown in Figure 6.6. The large number of pins make this board very handy for projects that require a bunch of digital inputs or outputs (like lots of LEDs or buttons).

Figure 6.6 Arduino Mega (R3).

6.4.2.7.2 Arduino Leonardo

The Leonardo is Arduino's first development board to use one microcontroller with built-in USB as shown in Figure 6.7. This means that it can be cheaper and simpler. Also, because the board is handling USB directly, code libraries are available which allow the board to emulate a computer keyboard, mouse, and more!

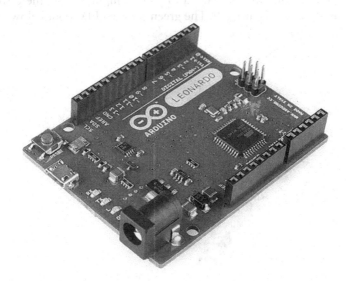

Figure 6.7 Arduino Leonardo.

How to set up the Arduino IDE on our computer:

Step 1: First you must have your Arduino board and a USB cable as shown in Figure 6.8.

Figure 6.8 USB cable.

Step 2: Download Arduino IDE software.
Download Arduino IDE from the Arduino official website. After your file download is complete, unzip the file.

Step 3: Power up your board.
Connect the Arduino board to your computer using the USB cable as shown in Figure 6.9. The green power LED should glow.

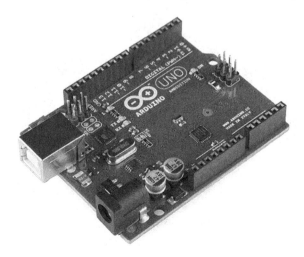

Figure 6.9 Arduino board.

Step 4: Launch Arduino IDE.

After your Arduino IDE software is downloaded, you need to unzip the folder. Inside the folder, you can find the application icon with an infinity label (application.exe).

Double-click the icon to start the IDE as shown in Figure 6.10.

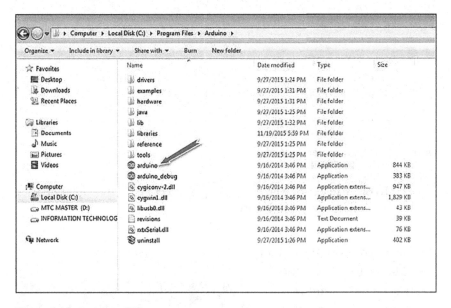

Figure 6.10 Arduino IDE.

Step 5: Open your first project as shown in Figure 6.11.

Once the software starts, you have two options:

- Create a new project.
- Open an existing project example.

To create a new project, select File → New.

⊛ sketch_jan18a | Arduino 1.8.13

File Edit Sketch Tools Help

sketch_jan18a

```
void setup() {
  // put your setup code here, to run once:

}

void loop() {
  // put your main code here, to run repeatedly:

}
```

Figure 6.11 Open project in Arduino.

To open an existing project example, select File → Example → Basics → Blink as shown in Figure 6.12.

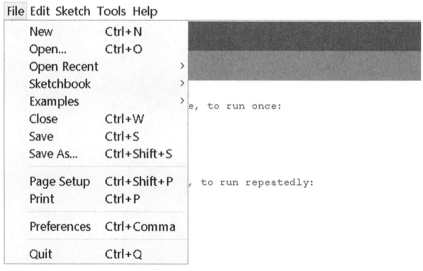

⊛ sketch_jan18a | Arduino 1.8.13

File Edit Sketch Tools Help

New	Ctrl+N
Open...	Ctrl+O
Open Recent	>
Sketchbook	>
Examples	>
Close	Ctrl+W
Save	Ctrl+S
Save As...	Ctrl+Shift+S
Page Setup	Ctrl+Shift+P
Print	Ctrl+P
Preferences	Ctrl+Comma
Quit	Ctrl+Q

Figure 6.12 How to open existing project.

Step 6: Select your Arduino board.

To avoid any error while uploading your program to the board, you must select the correct Arduino board name, which matches with the board connected to your computer.

Go to Tools → Board and select your board as shown in Figure 6.13.

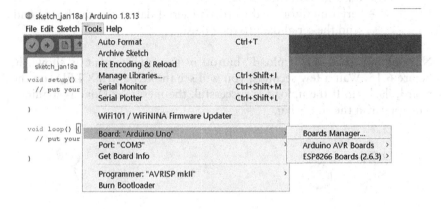

Figure 6.13 Select Arduino board.

Step 7: Select your serial port.

Select the serial device of the Arduino board.

Go to **Tools → Serial Port** menu.

This is likely to be COM3 or higher (COM1 and COM2 are usually reserved for hardware serial ports).

To find out, you can disconnect your Arduino board and re-open the menu, the entry that disappears should be of the Arduino board.

Reconnect the board and select that serial port as shown in Figure 6.14.

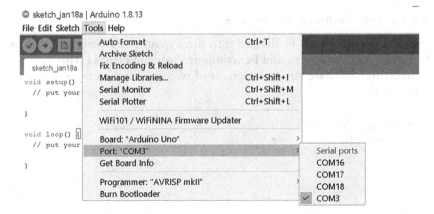

Figure 6.14 Serial port.

Step 8: Upload the program to your board.

   A – Used to check if there is any compilation error.
   B – Used to upload a program to the Arduino board.
   C – Shortcut used to create a new sketch.
   D – Used to directly open one of the example sketch.
   E – Used to save your sketch.
   F – Serial monitor used to receive serial data from the board and send the serial data to the board.

Now, simply click the "Upload" button in the environment as shown in Figure 6.15. Wait a few seconds; you will see the RX and TX LEDs on the board, flashing. If the upload is successful, the message "Done uploading" will appear in the status bar.

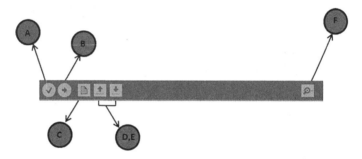

Figure 6.15 Upload program to board.

The Arduino programming language is a **modified version of C/C++**. Usually, we program in C++ with the addition of methods and functions. A program written in Arduino programming language is called **sketch** and saved with an **.ino** extension. You can even use **Python** to write an Arduino program. All these languages are text-based programming languages.

6.4.2.8 Arduino program structure

Arduino programs can be divided into three main parts: **Structure, Values** (variables and constants), and **Functions** as shown in Figure 6.16.

   Structure: Software structure consist of two main functions as shown in Figure 6.17:

- Setup( ) function
- Loop( ) function

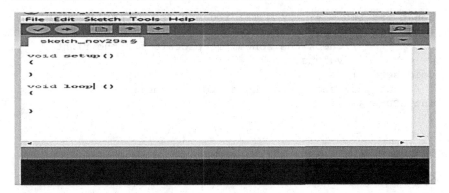

Figure 6.16 Arduino structure.

Other functions must be created outside the brackets of these two functions.

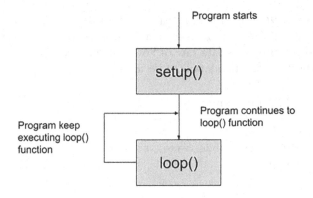

Figure 6.17 Structure of Arduino.

Structure
StructureVoid setup()
{
}

PURPOSE – The **setup()** function is called when a sketch starts.
Use it to initialize the variables, pin modes, start using libraries etc.
The setup function will only run once, after each power up or reset of the Arduino board.

Macros

#define LED 4
`--or--`
`    // Variables`

```
const int LED = 4;   // NOTE! You can use either 4 or D2
value for the same pin.
            //Check the schematic picture for the right
               values.
void setup()
{ // put your setup code here, to run once:
int LED = 4;
}
```

```
Serial.begin()
```

Serial.begin(): Sets the data rate in bits per second (baud) for serial data transmission.

This starts serial communication, so that the Arduino can send out commands through the USB connection.

Serial.begin(9600): passes the value 9600 to the speed parameter.

- The value 9600 is called the 'baud rate' of the connection.
- This tells the Arduino to get ready to exchange messages with the Serial Monitor at a data rate of 9600 bits per second.
- 9600 is how fast the data is to be sent.

Serial.begin() Function Syntax

```
void setup()
{ // put your setup code here, to run once:
Serial.begin(9600);
}
```

Arduino - I/O Functions

The pinMode() function is used to configure a specific pin to behave either as an input or as an output.

Pins

- The pins on the Arduino board can be configured as either inputs or outputs.
- Arduino pins are by default configured as inputs

```
pinMode() Function
 Void setup ()
 {
 pinMode (pin, mode);
 }
```

- pin – the number of the pin whose mode you wish to set
- mode – INPUT, OUTPUT,

6.4.3 Raspberry Pi

This is a fully functional low-cost credit-card sized computer, using Linux, which can be plugged into a monitor or TV as shown in Figure 6.18. To use it, simply connect it with a USB phone charger to power it, plug in a mouse and keyboard, and connect it to a TV and monitor together with an SD card stored with an operating system just like the one in your Phone. The Raspberry Pi is able to control and interact with electronic components such as sensors and actuators and explore the Internet of Things (IoT). It is capable of doing everything you'd expect a desktop computer to do, from browsing the Internet and playing high-definition video, to making spreadsheets, word processing and playing games.

Figure 6.18 Raspberry Pi.

These computers are a series of single board computers made by the Raspberry-Pi foundation, a UK charity that aims to educate people in computing and create easier access to computing education. The Raspberry Pi launched in 2012 and a number of variations have been released since then. All over the world people use Raspberry Pi to learn programming skills, build hardware projects, do home automation etc.

6.4.4 What Raspberry Pi models have been released?

There have been many generations of the Raspberry Pi line: from Pi 1 to 4, and even a Pi 400. There has generally been a Model A and a Model B of most generations. Model A has been a less expensive variant, and tends to have reduced RAM and fewer ports (such as USB and Ethernet). The Pi Zero is a spinoff of the original (Pi 1) generation, made even smaller and cheaper. Here's the lineup so far:

- Pi 1 Model B (2012)
- Pi 1 Model A (2013)
- Pi 1 Model B+ (2014)
- Pi 1 Model A+ (2014)
- Pi 2 Model B (2015)
- Pi Zero (2015)
- Pi 3 Model B (2016)
- Pi Zero W (2017)
- Pi 3 Model B+ (2018)
- Pi 3 Model A+ (2019)
- Pi 4 Model A (2019)
- Pi 4 Model B (2020)
- Pi 400 (2021)

6.4.5 Top 6 models of Raspberry Pi

The most notable models of Raspberry Pi available on the market are:

6.4.5.1 Raspberry Pi Zero

This is the cheapest Raspberry model produced by the company and can be purchased for as little as $5, which is quite impressive considering the extent of its functionality. Although not the first model to be released, it boasts a smaller, more compact size than the Raspberry Pi 1. Raspberry Pi Zero has the same processor and RAM (512 MB) as the Pi 1 Model B+. The Raspberry Pi Zero does not come with Wi-Fi or Bluetooth, but it can be made Internet-accessible via USB.

Its slightly more expensive version, Raspberry Pi Zero W, comes with Bluetooth 4.0 and a built-in 802.11n Wi-Fi connectivity. For projects that require GPIO pins, other versions of Raspberry Pi may be more suitable.

6.4.5.2 Raspberry Pi 1

Raspberry Pi 1 Model B was released in 2012. It served as a baseline in size for future releases. Initially, it had 26 GPIO pins, 256MB RAM capacity, and a single CPU core. You couldn't use it for heavy tasks with high processing needs. In 2014, the Raspberry Pi B+ was released with a starting RAM capacity of 512MB and 40 GPIO pins, becoming standard across all other models. Raspberry Pi Model B+ is sold at $25 and comes with four USB ports and an Ethernet connection. Pi 1 Model A+ ($20) can be considered for faster CPU processing speed, but it comes without an Ethernet connection.

6.4.5.3 Raspberry Pi 2 B

In February 2015, Raspberry released the 2B model. Compared to the prior releases, Raspberry Pi 2 B improved significantly, specifically in terms of memory and speed. The RAM capacity was increased to 1GB. Pi 2B comes in the standard size, with 4 USB ports. It is currently priced at about $35, which is pretty affordable.

6.4.5.4 Raspberry Pi 3

Raspberry Pi 3 B was released in 2016. The B+ version, which came out in 2018, can boast a faster processing unit, Ethernet (802.11ac), and Wi-Fi than the earlier version. Generally, Raspberry PI 3 offers the user a wide range of uses. It comes with the standard HDMI and USB ports, 1GB RAM, and Wi-Fi and Bluetooth connections in addition to the already functional Ethernet. One remarkable thing about this model is that it doesn't generate much heat or consume too much power. This makes it suitable for projects that require passive cooling and can be acquired at $35.

6.4.5.5 Raspberry Pi 4B

Released in 2019, the Raspberry Pi 4B is a vast improvement on its predecessors, with a varying memory capacity from 2GB RAM to 8GB RAM. It also has a faster 1.5GHz processor and a good mix of 2.0 and 3.0 USB ports. Pi 4B is an ideal Raspberry model as it is suitable for virtually every use case, having a higher RAM capacity to satisfy even the most dedicated programmers. Depending on memory, the price ranges from $35 to $75, but each comes with all connectivity options.

6.4.5.6 Raspberry Pi 400

This model is unique as it comes in the form of a keyboard. It was launched in 2020 and operated with 4GB RAM. It comes with standard USB ports and requires just a monitor and a mouse to make it a home computer set. Pi 400 costs $70 and can be used effectively in classrooms.

Out of the above versions of Raspberry Pi, more prominently use Raspberry Pi and their features are as follows in Table 6.1.

Table 6.1 Raspberry Pi versions

Features	Raspberry Pi Model B+	Raspberry Pi 2 Model B	Raspberry Pi 3 Model B	Raspberry Pi zero
SoC	BCM2835	BCM2836	BCM2837	BCM2835
CPU	ARM11	Quad Cortex A7	Quad Cortex A53	ARM11
Operating Freq.	700 MHz	900 MHz	1.2 GHz	1 GHz
RAM	512 MB SDRAM	1 GB SDRAM	1 GB SDRAM	512 MB SDRAM
GPU	250 MHz Videocore IV	250MHz Videocore IV	400 MHz Videocore IV	250MHz Videocore IV
Storage	micro-SD	micro-SD	micro-SD	micro-SD
Ethernet	Yes	Yes	Yes	No
Wireless	WiFi and Bluetooth	No	No	No

6.4.6 How to install Raspbian on the Raspberry Pi

The first step in configuring the Raspberry Pi would be to install the Raspbian operating system. Go to https://www.raspberrypi.org/downloads/ and select the Raspbian OS as shown in Figure 6.19.

Figure 6.19 How to install Raspbian OS.

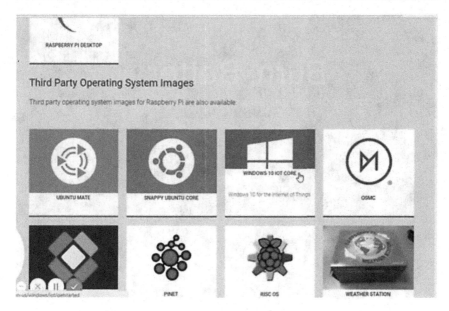

Figure 6.20 Install operating system in Raspberry Pi.

- Once you download the Raspbian operating system, you would need to format the SD card and setup the Raspbian OS onto your SD card as shown in Figure 6.20 and Figure 6.21.
- After the download is finished, you will end up with a zip file. Unzipping this, you will get an image file and you just write that image onto the memory card as shown in Figure 6.22 and Figure 6.23.

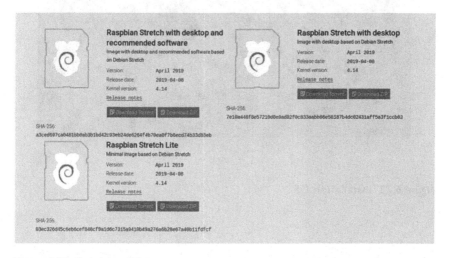

Figure 6.21 Raspbian OS.

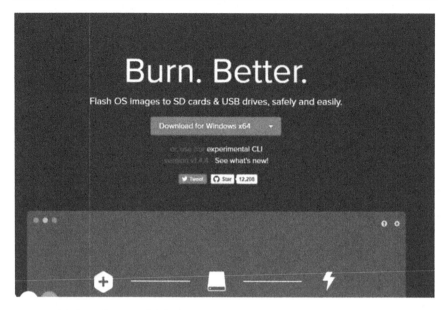

Figure 6.22 Installation of OS.

Figure 6.23 Installation OS.

Figure 6.24 Installed OS.

- For Windows-based systems, which is used by most people, the installation for this image is quite easy as shown in Figure 6.24.

It requires a piece of software called Win32 Disk Imager as shown in Figure 6.25.

- Download the Win32DiskImager ZIP file.
- Expand the ZIP file to a folder on disk.
- Download a Raspberry Pi distribution disk image.
- Run Win32DiskImager.exe from the install folder.
- Select the source image file and the target device.
- Click on the **Write** button to copy the image to the SD card.
- Writing an image to a disk takes about 5 minutes for a 2-GB image file. Once the image is written to the SD card, the SD card may be ejected and used to boot the Raspberry Pi as shown in Figures 6.26–6.29.

🌣 **Win32 Disk Imager**			_ □ ✕
Image File			Device
|		📂	⌄
Copy ☐ MD5 Hash:			
Progress			
Version: 0.9	Cancel	Read	Write	Exit

Figure 6.25 Win32 Disk Manager.

Figure 6.26 SD card.

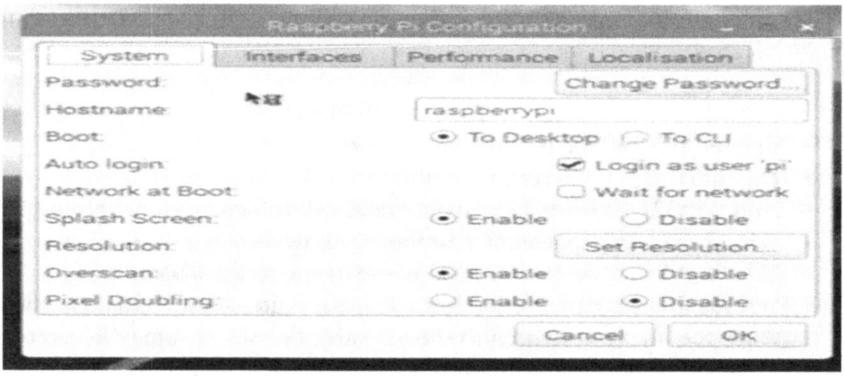

Figure 6.27 Raspberry Pi installation settings.

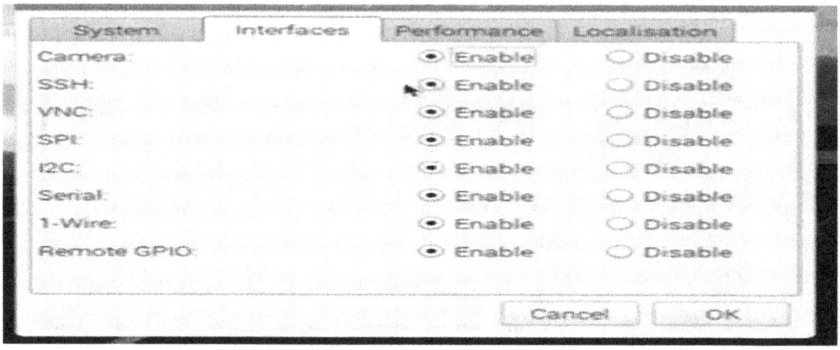

Figure 6.28 Interfaces in Raspberry Pi.

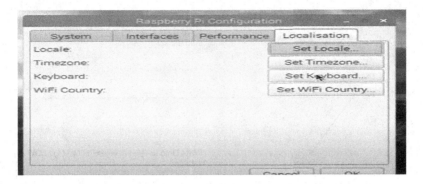

Figure 6.29 Localization in Raspberry Pi.

6.4.7 Raspberry Pi pins

The best thing about any Raspberry Pi, including the new Raspberry Pi 4, is that you can use it to build all kinds of awesome contraptions, from robots to retro gaming consoles (and even fart detectors!). Most of the sensors, motors, lights and other peripherals that make these projects possible connect to the Pi's set of GPIO (General Purpose Input Output) pins. These pins offer a direct connection to the System on Chip (SoC) at the heart of the Pi, enabling the Pi to communicate with external components as shown in Figure 6.30. Every Pi model since the Raspberry Pi B+ has had 40 GPIO pins, though on the Pi Zero and Zero W, you have 40 holes into which you can solder pins or wires as shown in Figure 6.31. If you haven't got a soldering iron, fear not, we have a list of the best soldering irons for you to choose from.

This guide has been updated to reflect the new capabilities of the Raspberry Pi 4, which still comes with 40 GPIO pins, but has a few extra I2C, SPI, and UART connections available.

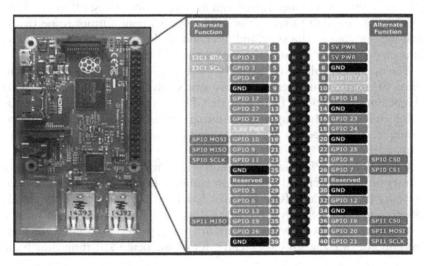

Figure 6.30 Pin layout in Raspberry Pi.

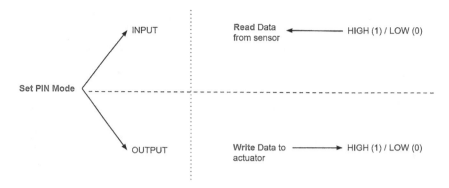

Figure 6.31 PIN mode in Raspberry Pi.

1. Ground pins
2. Power pins
3. Reserved pins
4. Raspberry Pi GPIOs
   - GPIOs are digital pins
   - GPIOs voltage
   - To use a GPIO, first you need to know its number
   - The pin numbers and GPIO numbers are different
   - Pin numbers are in grey, and GPIO numbers in orange
   - Depending on the library you use to manipulate GPIOs, you'll either have to use the number of the pin or the GPIO number

6.4.8 General Purpose Input Output (GPIO) Pins

The GPIO is the most basic, yet accessible aspect of the Raspberry Pi. GPIO pins are digital which means they can have only two states: off or on. They can have a direction to receive or send current (input, output respectively) and we can control the state and direction of the pins using programming languages such as Python, JavaScript, node-RED etc.

The operating voltage of the GPIO pins is 3.3v with a maximum current draw of 16mA. This means that we can safely power one or two LEDs (Light Emitting Diodes) from a single GPIO pin, via a resistor (see resistor color codes). But for anything requiring more current, a DC motor, for example, we will need to use external components to ensure that we do not damage the GPIO.

Controlling a GPIO pin with Python is accomplished by first importing a library of pre-written codes. The most common library is RPi.GPIO (https://pypi.org/project/RPi.GPIO/) and this has been used to create thousands of projects since the early days of the Raspberry Pi. In more recent times a new library called GPIO Zero (https://pypi.org/project/gpiozero/) has been introduced, offering an easier entry for those new to Python and basic electronics. Both of these libraries come pre-installed with the Raspbian operating system.

GPIO pins have multiple names; the first most obvious reference is their "physical" location on the GPIO. Starting at the top left of the GPIO, and by that we mean the pin nearest to where the micro SD card is inserted, we have physical pin 1, which provides 3v3 power. To the right of that pin is physical pin 2, which provides 5v power. The pin numbers then increase as we move down each column, with pin 1 going to pin 3, 5,7 etc. until we reach pin 39. You will quickly see that each pin from 1 to 39 in this column follows an odd number sequence. And for the column starting with pin 2 it will go 4,6,8 etc. until it reaches 40,following an even number sequence. Physical pin numbering is the most basic way to locate a pin, but many of the tutorials written for the Raspberry Pi follow a different numbering sequence.

Broadcom (BCM) pin numbering (aka GPIO pin numbering) seems to be chaotic to the average user. With GPIO17, 22 and 27 following on from each other with little thought to logical numbering. The BCM pin mapping refers to the GPIO pins that have been directly connected to the System on a Chip (SoC) of the Raspberry Pi. In essence, we have direct links to the brain of our Pi to connect sensors and components for use in our projects.

You will see the majority of Raspberry Pi tutorials using this reference and that is because it is the officially supported pin numbering scheme from the Raspberry Pi Foundation. So it is best practice to start using and learning the BCM pin numbering scheme as it will become second nature to you over time. Also note that BCM and GPIO pin numbering refer to the same scheme. So, for example, GPIO17 is the same as BCM17.

Certain GPIO pins also have alternate functions that allow them to interface with different kinds of devices that use the I2C, SPI, or UART protocols. For example, GPIO3 and GPIO 4 are also SDA and SCL I2C pins used to connect devices using the I2C protocol. To use these pins with these protocols we need to enable the interfaces using the Raspberry Pi Configuration application found in the Raspbian OS, Preferences menu.

6.4.9 I2C, SPI, and UART: Which do you use?

We'll get into the specific differences between I2C, SPI, and UART below, but if you're wondering which one you need to use to connect to a given device, the short answer is to check the spec sheet. For example, one tiny LED screen might require SPI and another might use I2C (almost nothing uses UART). If you read the documentation that comes with a product (provided it has some), it will usually tell you which Pi pins to use.

For Raspberry Pi 4 users, note that there are now many more I2C, SPI, and UART pins available to you. These extra interfaces are activated using device tree overlays and can provide four extra SPI, I2C, and UART connections.

6.4.10 I2C – Inter-Integrated Circuit

I2C is a low-speed two-wire serial protocol to connect devices using the I2C standard. Devices using the I2C standard have a master–slave relationship. There can be more than one master, but each slave device requires a unique address, obtained by the manufacturer from NXP, formerly known as Philips Semiconductors. This means that we can talk to multiple devices on a single I2C connection as each device is unique and discoverable by the user and the computer using Linux commands such as i2cdetect.

As mentioned earlier, I2C has two connections: SDA and SCL. They work by sending data to and from the SDA connection, with the speed controlled via the SCL pin. I2C is a quick and easy way to add many different components, such as LCD/OLED screens, temperature sensors, and analog to digital converters for use with photoresistors etc. to your project. While proving to be a little more tricky to understand than standard GPIO pins, the knowledge gained from learning I2C will serve you well as you will understand how to connect higher-precision sensors for use in the field.

The Raspberry Pi has two I2C connections at GPIO 2 and 3 (SDA and SCL) are for I2C0 (master) and physical pins 27 and 28 are I2C pins that enable the Pi to talk to compatible HAT (Hardware Attached on Top) add-on boards.

6.4.11 SPI – Serial Peripheral Interface

Serial Peripheral Interface (SPI) is another protocol for connecting compatible devices to your Raspberry Pi. It is similar to I2C in that there is a master–slave relationship between the Raspberry Pi and the devices connected to it.

Typically, SPI is used to send data over short distances between microcontrollers and components such as shift registers, sensors, and even an SD card. Data is synchronized using a clock (SCLK at GPIO11) from the master (our Pi) and the data is sent from the Pi to our SPI component using the MOSI (GPIO GPIO10) pin. MOSI stands for Master Out Slave In. If the component needs to reply to our Pi, then it will send data back using the MISO pin (GPIO9) which stands for Master In Slave Out.

6.4.12 UART – Universal Asynchronous Receiver/Transmitter

Commonly known as "Serial," the UART pins (Transmit GPIO14, Receive GPIO15) provide a console/terminal login for headless setup, which means connecting to the Pi without a keyboard or pointing device. Normally, the easiest way to do a headless Raspberry Pi setup is simply to control the Pi over a network or direct USB connection (in the case of Pi Zero).

If there's no network connection, however, you can also control a headless Pi using a serial cable or USB to serial board from a computer running a terminal console. UART is exceptionally reliable and provides access to a Pi without the need for extra equipment. Just remember to enable the Serial Console in the Raspberry Pi Configuration application. Chances are that you won't want to do this, but the UART support is there if you need it.

6.4.13 Ground (gnd)

Ground is commonly referred to as GND, gnd or – but they all mean the same thing. GND is where all voltages can be measured from, and it also completes an electrical circuit. It is our zero point and by connecting a component, such as an LED, to a power source and ground the component becomes part of the circuit and current will flow through the LED and produce light.

When building circuits it is always wise to make your ground connections first before applying any power as it will prevent any issues with sensitive components. The Raspberry Pi has eight ground connections along the GPIO and each of these ground pins connects to one single ground connection. So the choice of which ground pin to use is determined by personal preference, or convenience when connecting components.

6.4.14 5v

The 5V pins give direct access to the 5V supply coming from your mains adaptor, less power than used by the Raspberry Pi itself. A Pi can be powered directly from these pins, and it can also power other 5V devices. When using these pins directly, be careful and check your voltages before making a connection because they bypass any safety features, such as the voltage regulator and fuse which are there to protect your Pi. Bypass these with a higher voltage and you could render your Pi inoperable.

6.4.15 3v3

The 3V pin is there to offer a stable 3.3V supply to power components and to test LEDs. In reality, it will be rare that you factor this pin into a build, but it does have a special use. When connecting an LED to the GPIO, we first need to make sure that the LED is wired up correctly and that it lights up. By connecting the long leg of the LED, the anode to the 3.3V pin via a resistor, and the shorter leg, the cathode, to any of the Ground (gnd) pins we can check that our LED lights up and is working. This eliminates a hardware fault from the project and enables us to start building our project with confidence.

Communication protocols through Raspberry Pi 4 pins:

- UART is a multi-master communication protocol. This protocol is quite easy to use and very convenient for communicating between several boards: Raspberry Pi to Raspberry Pi, or Raspberry Pi to Arduino, etc.

To use UART you need three pins:

- GND that you'll connect to the global GND of your circuit.
- RX for Reception. You'll connect this pin to the TX pin of the other component.
- TX for Transmission. You'll connect this pin to the RX of the other component.

- I2C is a master–slave bus protocol (well it can have multiple masters, but you'll mostly use it with one master and multiple slaves). The most common use of I2C is to read data from sensors and actuate some components.

- SPI is yet another hardware communication protocol. It is a master–slave bus protocol. It requires more wires than I2C, but can be configured to run faster.

For using SPI you'll need 5 pins:

- GND: what a surprise! Make sure you connect all GND from all your slave components and the Raspberry Pi together.
- SCLK: clock of the SPI. Connect all SCLK pins together.
- MOSI: means Master Out Slave In. This is the pin to send data from the master to a slave.
- MISO: means Master In Slave Out. This is the pin to receive data from a slave to the master.
- CS: means Chip Select. Pay attention here: you'll need one CS per slave on your circuit. By default you have two CS pins (CS0 – GPIO 8 and CS1 – GPIO 7). You can configure more CS pins from the other available GPIOs.

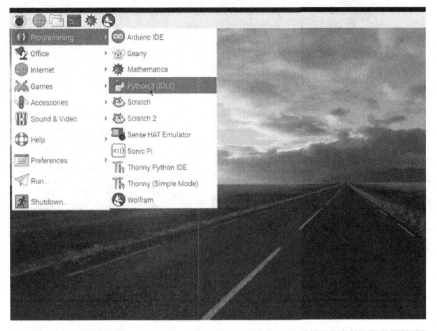

```
Python 3.5.3 (default, Jan 19 2017, 14:11:04)
[GCC 6.3.0 20170124] on linux
Type "copyright", "credits" or "license()" for more information.
>>>
                    I
```

Figure 6.32 Raspberry Pi installation.

Installed by default on Raspberry Pi as shown in Figure 6.32:

C
C++
Java
Scratch
Ruby

Program1: To on LED as shown in Figure 6.33:

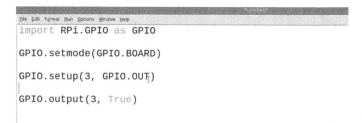

```
File  Edit  Format  Run  Options  Window  Help
import RPi.GPIO as GPIO

GPIO.setmode(GPIO.BOARD)

GPIO.setup(3, GPIO.OUT)

GPIO.output(3, True)
```

Figure 6.33 LED program.

Run-> Run Module
GPIO . setmode (GPIO . BCM) (as shown in Figure 6.34 and 6.35):

Figure 6.34 LED setup.

Figure 6.35 LED connectivity.

Program 2: To Blink LED as shown in Figure 6.36:

```
File Edit Format Run Options Window Help
import RPi.GPIO as GPIO
import time

GPIO.setmode(GPIO.BOARD)

GPIO.setup(3, GPIO.OUT)

while True:
      GPIO.output(3, True)
      time.sleep(1)
      GPIO.output(3, False)
      time.sleep(1)
```

Figure 6.36 Blink LED program.

Program 3: Brightness control of LED using PWM as shown in Figure 6.37:

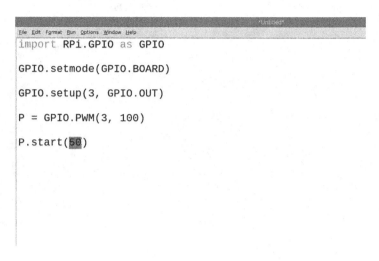

Figure 6.37 Brightness control of LED using PWM.

Program 4: Brightness gradually increases/decreases

- Use for loop:

P . start(0)
While True:
For X in range (100);
P . Start (X)
Time . sleep (0.1)
For X in range (100);
P . Start (100-X)
Time . sleep (0.1)

Chapter 7

Security aspects in IoT

Every connected device creates opportunities for attackers. These vulnerabilities are broad, even for a single small device. The risks posed include data transfer, device access, malfunctioning devices, and always-on/always-connected devices. The main challenges in security remain the security limitations associated with producing low-cost devices, and the growing number of devices, which creates more opportunities for attacks.

7.1 SECURITY SPECTRUM

The definition of a secured device spans from the simplest measures to sophisticated designs. Security should be thought of as a spectrum of vulnerability which changes over time as threats evolve. Security must be assessed based on user needs and implementation. Users must recognize the impact of security measures because poorly designed security creates more problems than it solves.

Example: A German report revealed hackers compromised the security system of a steel mill. They disrupted the control systems, which prevented a blast furnace from being shut down properly, resulting in massive damage. Therefore, users must understand the impact of an attack before deciding on the appropriate level of protection (Figure 7.1).

The three main point's attackers can access IoT devices connected to a network are:

1. The device
2. The cloud
3. The network
   - **Securing the Device:** There are some technologies in the industry such as embedded SIM Technology (eUICC), M2M-optimized SIM Technology, SafeNet Hardware Security Modules (HSMs), Trusted Key Manager, and IP protection to provide security for embedded

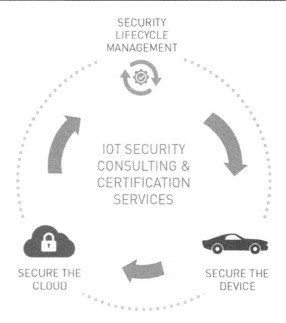

Figure 7.1 IoT services.

devices. My opinion is that IP protection is a little bit out of date. Current IoT ecosystems should move from such security infrastructures to something more advanced with encryption technologies.

- **Securing the cloud infrastructure**: A major form of threat comes from the enterprise or cloud environment to which smart devices are connected. Data encryption, cloud security, and cloud-based licensing helps technology companies leverage the full potential of the cloud environment, ensuring their intellectual property is secured.
- **IoT security lifecycle management**: Managing the lifecycle of security components across the device and cloud spectrum is a critical element for a robust and long-term digital security strategy. The security of an Internet of ecosystems is not a one-off activity, but an evolving part of the Internet of ecosystems. Among the proposed solutions suggested for building a sustainable security lifecycle management infrastructure to address current and future security threats are identity and access management, crypto management, and maintaining trusted services hubs (Figure 7.2).

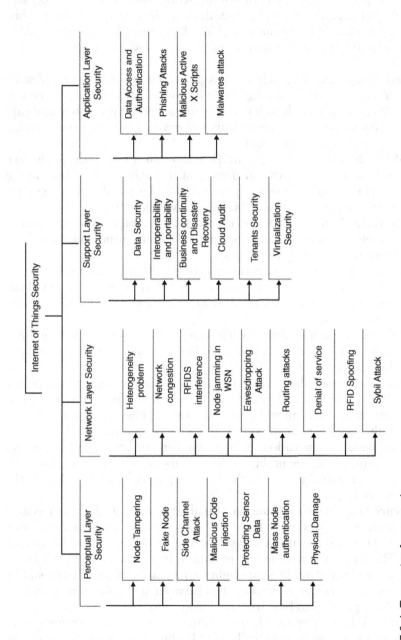

Figure 7.2 IoT security framework.

7.2 PERCEPTUAL LAYER SECURITY

The perceptual layer consists of resource-constrained IoT devices i.e. sensors, RFID tags, Bluetooth and Zigbee devices. These devices are more prone to cyber-attacks. As a large amount of IoT devices are physically deployed in open fields, they encounter many physical attacks, which include:

Node tampering: If an attacker has physical access to sensor nodes, he or she can replace the full node or part of its hardware. They can also connect directly to it to alter some sensitive information and gain access to the node. The sensitive information may be cryptographic keys or routing table's routes.

Fake node attacker: A cyber-attacker can add a fake node to the IoT system and can inject malicious data through this fake node in the network, thus making low-power devices busy and consuming their energy. It can also act as a Man in the Middle attack.

Side channel attack: Attackers use the information such as power consumption, time consumption, and electromagnetic radiation from senor nodes to attack encryption mechanisms.

Physical damage: The adversary can physically damage the IoT device for denial of service (DoS) purposes. IoT devices are deployed in both open and closed vicinities and are more susceptible to physical damage by the attacker.

Malicious code injection: An adversary can physically compromise a node by inserting malicious code into the node that will give them illegal access to the system.

Protecting sensor data: The confidentiality requirements of the sensor data is low as an adversary can place a sensor near to the IoT system sensor and can sense the same value. However, its integrity and authenticity is more important and must be secured.

Mass node authentication: A large number of nodes in an IoT system face authentication problems. A huge amount of network communication is required just for authentication purpose, thereby affecting the performance.

Security requirements of the perceptual layer: First of all, the IoT system must be physically secured from its adversary gaining access. Node authentication is also necessary to prevent illegal access to the system. The integrity confidentiality of data to be transmitted between nodes is very important, so lightweight cryptographic algorithms should be designed to securely transmit data between nodes. Key management is also a problem to be solved in context of IoT.

Network layer security: The core network has sufficient security measures but certain issues still exist. Traditional security problems can affect the integrity and confidentiality of data. Many types of network attacks, such as an eavesdropping attack, a DoS attack, a Man in the Middle attack, and virus invasion are still affecting the network layer.

Heterogeneity problem: The IoT perceptual layer is the combination of many heterogeneous technologies. The access network has multiple access methods, and this heterogeneity makes security and interoperability more challenging.

Network congestion problems: A large amount of sensor data, along with the communication overheads caused by the presence of a large number of devices to be authenticated, can cause network congestion. This problem should be solved by introducing a feasible device authentication mechanism and competent transport protocols.

RFIDs interference: This is an attack on the network layer in which the radio frequency signals used by RFIDs are corrupted with noise signals, thereby causing Denial of Service.

Node jamming in WSN: This is a similar type of attack to radio frequency interference as discussed above for RFIDs. In this attack the attacker interferes with the radio frequency of wireless sensor networks and denies services from WSNs. It is also a type of Denial of Service attack.

Eavesdropping attack: This involves the sniffing out of traffic in the wireless vicinity of WSNs, RFIDs or Bluetooth due to the wireless nature of the device layer in IoT. Every type of attack starts from information gathering via sniffing using available tools such as packet sniffers.

Denial of Service: In this attack the adversary overburdens the network by swamping it with traffic above its capacity and thus the network is unavailable for useful services to legitimate users.

RFID spoofing: The attacker initially sends spoof RFID signals and read RFID tags. The attacker then sends fake data with the original RFID tag and gains full access to the system.

Routing attacks: The adversary can alter the routing information and distribute it in the network to create routing loops, advertising false routes, sending error messages, or dropping network traffic.

Sybil attack: In a Sybil attack, a single malicious node claims the identity of many nodes. This node can cause much damage; it can distribute false routing information or it can also rag the WSN selection process.

Security requirements of the network layer: Although the existing core network security is mature enough, some security concerns still exist which are more harmful in the context of IoT, such as Denial of Service and Distributed Denial of Service attacks, must be prevented in this layer. Communication protocols must be very mature to solve the problem of routing attack, congestion problems, and spoofing security problems.

Support layer Security: Support layer security is independent from other layers and cloud computing security is a large domain of security. The Cloud Security Alliance (CSA) is setting many standard security frameworks for clouds. It is also developing mechanism for continuous cloud audits, such as Security Content Automation Protocol (SCAP) and providing trusted results via Trusted Computing (TCG). This layer hosts the IoT user's data and applications, so both should be protected from security breaches. Among the security concerns at this layer are:

Data security: To keep the data confidential and secure in the cloud it must be secure from breaches. This can be done by using tools to detect data migration from cloud, data loss prevention tools, and file and database activity monitoring. Data dispersion and data fragmentation can also be used for data security in the cloud.

Interoperability and portability: Interoperability and portability among cloud vendors is a major present-day problem. Different vendors use different proprietary standards, creating problems for users who want to migrate from one cloud to another. This heterogeneity also create security exposure.

Business continuity and disaster recovery: Cloud vendors must provide a continuity of services in natural disasters such as floods, fires, and earthquakes. In order to achieve this, the cloud's physical location should be suitable so that it is the least affected by such calamities. It should be in the approach of quick-response teams. Clouds should also have some data backup plans.

Cloud audit: The Cloud Security Alliance sets many standards for cloud vendors. A continuous audit is required to check the compliance of these security standards to build user trust.

Tenants Security: The data of multiple users may be located at same physical drive in the cloud or users of Infrastructure as a Service (IaaS) may share the same physical storage; such users are called tenants. The adversary can steal his/her tenant's data as the data share the same physical media.

Virtualization security: Different cloud vendors used different virtualization techniques. The security of virtualization is important. Virtual machine communication can occasionally bypass network security controls. The secure migration of a virtual machine is required as it can be a hurdle in cloud audit.

Security Requirements of Support Layer: Internet of Things user data and application instances resides on cloud and fog nodes. There security and privacy should not be abused in the cloud. The Cloud Security Alliance (CSA) has already set many security standards, laws, and regulations for cloud security. The compliance of these security standards should be monitored continuously and IoT systems should only use those clouds which comply with the security standards of CSA.

Application layer security: Different applications at the application layer have different security requirements. By now there is no standard for IoT application construction. However, data sharing is one of the characteristics of the IoT application layer. Data sharing face problems of data privacy and access control. Some of the common security matters of application layer are:

Data access and authentication: An application may have many users, and different users may have different access privileges. Proper authentication and access control mechanism is required at the application layer.

Phishing attacks: The adversary uses infected emails or web links to steal legitimate user credentials and gain access using those credentials

Malicious Active X scripts: The adversary can send an Active X script to the IoT user through the Internet and make the IoT user to run the Active X script, thus compromising the whole system.

Malware attacks: An attacker can attack applications using malware and can steal data or cause Denial of Service. Trojan horses, worms, and viruses are among the dangerous malware used by adversaries to exploit a system.

Security requirements of the application layer: To cope with the application layer security, strong authentication and access control mechanism is required. Besides these educating the users to use a strong password is also important. Strong anti-virus software is required to protect against malware.

Security Threats in Smart Home: Smart home services can be exposed to cyber-attacks because the majority of service providers do not consider security parameters during the early stages of planning. The possible security threats in a smart home are eavesdropping, Distributed Denial of Service (DDoS) attacks, and the leakage of information, etc. Smart home networks are threatened by unauthorized access. The possible security threats to smart home are discussed as follows (see Figure 7.3).

- **Trespass:** If the smart door lock is effected by malicious codes or it is accessed by an unauthorized party, the attacker can trespass into a smart home without smashing the doorway (Figure 7.4). The result of this effect could be in the form of loss of life or property. To get rid of such attacks, passwords should be changed frequently. These should contain at least ten characters because it is very difficult for attackers

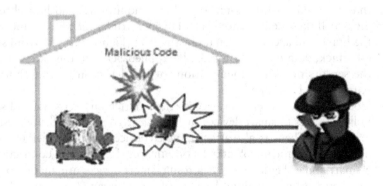

Figure 7.3 Security threats in smart home.

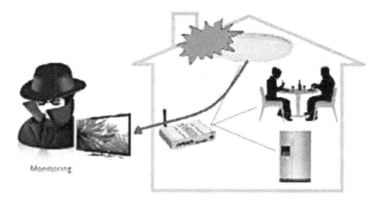

Figure 7.4 Security threats in smart home.

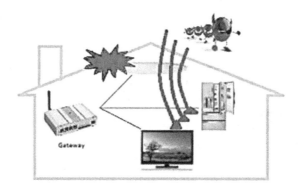

Figure 7.5 Security threats in smart home.

to break the long password. Similarly, authentication mechanism and access control may also be applied.

- **Monitoring and personal information leakage**: Safety is one of the important purposes of a smart home. Hence there are a lot of sensors that are used for fire monitoring, baby monitoring, and housebreaking, etc. If these sensors are hacked by an intruder then he can monitor the home and access personal information (Figure 7.5). To avoid such an attack, data encryption must be applied between the gateway and the sensors or user authentication for the detection of unauthorized parties may be applied.

- **DoS/DDoS**: Attackers may access the smart home network and send bulk messages to smart devices such as Clear To Send (CTS)/Request To Send (RTS). They can also attack targeted device by using malicious codes in order to perform DoS attacks on other devices that are connected in a smart home (Figure 7.6). As a result, smart devices are unable to perform proper functionalities because of the draining of resources due to

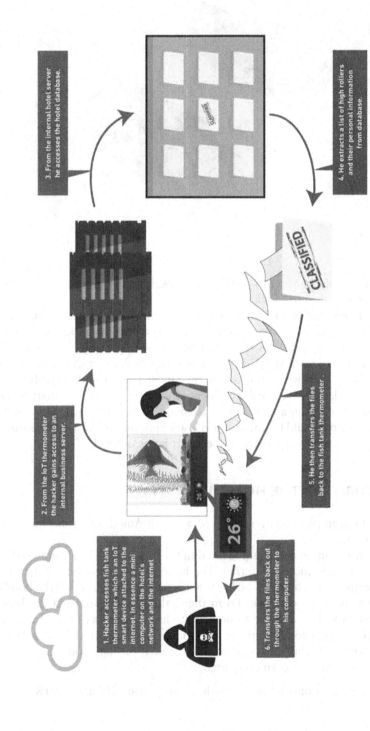

Figure 7.6 Security in the hotel industry.

Figure 7.7 USB.

such attacks. In order to avoid this attack, it is very important to apply authentication to block and detect unauthorized access.

- **Falsification:** When the devices in smart home perform communication with the application server, the attacker may collect the packets by changing the routing table in the gateway as shown in Figure 7.7. Although the secure socket layer (SSL) technique is applied, an attacker can bypass the forged certificate. In this way, the attacker can misinterpret the contents of data or may leak confidential data. To secure the smart home network from this attack, an SSL technique with a proper authentication mechanism should be applied. It is also important to block unauthorized devices that may try to access the smart home network.

7.3 SECURITY IN THE HOTEL INDUSTRY

Real-world example: Hackers attacked a North American casino hotel via an aquarium.

A thermostat in an aquarium located in the hotel lobby was connected to the hotel's servers and the Internet. The hackers compromised the thermostat and gained control of it. The hackers then found and accessed the hotel's servers via the hotel network. From there, the hackers attacked the hotel server and moved from there to breach the hotel database. From the database, hackers extracted high-roller information and exported the data back to the thermostat. The hackers then downloaded the high-roller data from the thermostat to their own computer.

Are these hotel rooms smart enough to keep you safe and protect your privacy?

Protecting hotel infrastructure against IoT vulnerabilities. IoT devices: CCTV, HVAC, electronic key card systems, fire detection. All of these devices can be hacked. CCTV security cameras could be hacked. A real-world example: In 2017 a four-star hotel in Austria was targeted by hackers (Phishing Email).

What hotels can do to mitigate risk:

- Educate employees
- Change the default usernames and passwords on all IoT devices.
- Not have unauthorized access to networked computers(technicians or maintenance teams),
- Create discrete, firewalled networks that separate IoT devices from hotel business, guest and visitor Wi-Fi.
  Protecting hotel guest room privacy against IoT vulnerabilities: Hacking guest phones through USB chargers. A real-world example: In 2015, during talks on the Iran nuclear deal at a five-star hotel in Geneva.

What guests and hotels can do to mitigate risk:

- Turning off smart TVs is not enough to ensure privacy.
- Do not use the room's USB ports and standard cables to charge your phone or tablet.
- Hotels should place all guest room IoT devices on a network separate from the hotel server.
- Smart TVs should be monitored by Intrusion Detection Systems.
- **Protecting hotel information systems against IoT vulnerabilities:** Hotel information systems contain everything from guest contact information and credit card numbers to hotel financial records, employee files, and security protocols.

What hotels can do to mitigate risk:

- Isolate all IoT devices
- Systems that need to be connected to the hotel's internal business network should be carefully set up by expert consultants
- Any IoT devices that do not require internet access should be isolated from the worldwide internet.

7.4 CASE STUDY: IP CAMERA

We shall consider the IP camera, how it is vulnerable, and what aspects of security needed to be considered. There are three components involved in this product: the camera itself, the controller, and cloud servers. Let us learn the aspects in which we need to look in for the security (Figure 7.8).

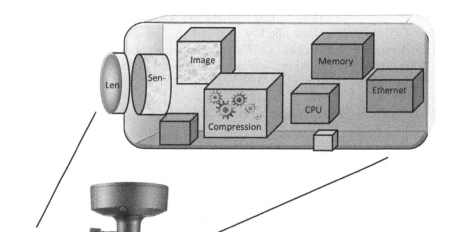

Figure 7.8 IP camera.

Microcontroller unit: It does not have a full-fledged operating system, but only a dedicated small piece of code, called firmware, which was written for a particular application.

- Code has to be carefully designed and implemented.
- Second component is the network.
- If the camera and the controller is placed in the unchanged network, then the controller can easily connect with the camera locally and send the video with the help of a web server on the camera.
- The Ethernet or Wi-Fi is used.
- The passwords can be changed along with other configurations, including the image resolution of the web page.

Three remote attacks against the IP camera of interest:

- device scanning attack
- brute force attack
- device spoofing attack.

Using these attacks, we can remotely control any camera.

Security architecture (Figure 7.9): Security architecture is the strategic design of systems, policies and technologies to protect IT and business assets

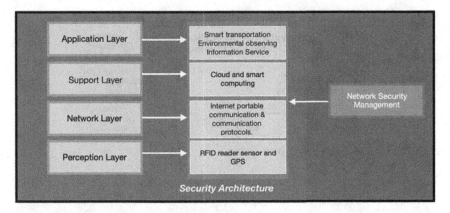

Figure 7.9 Security architecture.

from cyberthreats. A well-designed security architecture aligns cybersecurity with the unique business goals and risk management profile of the organization.

> **Perception layer (also called the recognition layer):** Gathers all type of information with the help of physical equipment. Physical equipment includes: RFID reader, sensors etc.
> **Network layer:** The second layer in the architecture is the network layer. It is responsible for the broadcasting of data. Data collected on numerous essential networks such as the mobile communication network, or the Wi-Fi network, satellite network, and more.
> **Support layer:** The third layer in the architecture is the support layer. It acts as a mediator. Grid and cloud computing are mostly used here.
> **Application layer:** The fourth layer in the architecture is called the application layer. The personalized delivery of application happens, whatever application the user wants, whatever application the user is presented with is taken care of in this layer.

7.5 IoT SECURITY TOOLS

- Encryption
- Password protection
- Two-factor authentication

> **Biometrics:** This is a lightweight cryptographic algorithm or protocol tailored for implementation in constrained environments, including RFID tags, sensors, contactless smart cards, healthcare devices, and so on. The lightweight primitives are superior to conventional cryptographic

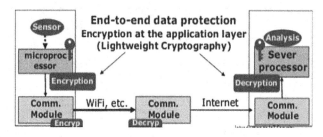

Figure 7.10 End-to-end data protection.

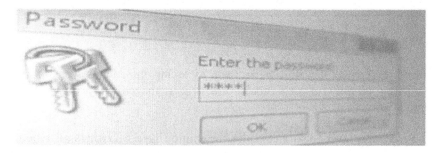

Figure 7.11 Password-based authentication.

ones, which are currently used in the Internet security protocols e.g. IPSec, TLS. It also delivers adequate security (Figure 7.10).

Password-based Authentication: This starts to look less attractive as a security solution for IoT devices, for two reasons: Passwords do not work well on dumb devices. They lack the power to process or store passwords. Passwords are a poor means of automated authentication. Entering a password generally requires a human to do something and that's hard to automate. As a result, passwords aren't good for securing automated exchanges of information (Figure 7.11).

Two-Factor Authentications (2FA): This is often referred as two-step verification. It is a security process in which the user provides two authentication factors to verify they are who they say they are. 2FA can be contrasted with single-factor authentication (SFA), a security process in which the user provides only one factor, typically a password. Two-factor authentication provides an additional layer of security and makes it harder for attackers to gain access. Two-factor authentication has long been used to control access to sensitive systems and data (Figure 7.12).

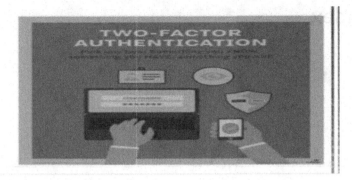

Figure 7.12 2-factor authentication.

Biometrics: This is the process of comparing data for the person's characteristics to that person's biometric template in order to determine resemblance. The reference model is first stored in a database or a secure, portable element like a smart card. The data stored is then compared to the person's biometric data in order for it to be authenticated.

Index

Printed in the United States
by Baker & Taylor Publisher Services